Selected Titles in This Series

∿ **W9-AYN-896**

Volume 10

CRM PROCEEDINGS & LECTURE NOTES

Centre de Recherches Mathématiques
Université de Montréal

Stable Marriage and Its Relation to Other Combinatorial Problems

An Introduction to the Mathematical Analysis of Algorithms

Donald E. Knuth

The Centre de Recherches Mathématiques (CRM) of the Université de Montréal was created in 1968 to promote research in pure and applied mathematics and related disciplines. Among its activities are special theme years, summer schools, workshops, postdoctoral programs, and publishing. The CRM is supported by the Université de Montréal, the Province of Québec (FCAR), and the Natural Sciences and Engineering Research Council of Canada. It is affiliated with the Institut des Sciences Mathématiques (ISM) of Montréal, whose constituent members are Concordia University, McGill University, the Université de Montréal, the Université du Québec à Montréal, and the Ecole Polytechnique.

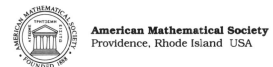

American Mathematical Society
Providence, Rhode Island USA

The production of this volume was supported in part by the Fonds pour la Formation de Chercheurs et l'Aide à la Recherche (Fonds FCAR) and the Natural Sciences and Engineering Research Council of Canada (NSERC).

Translated by Martin Goldstein

1991 *Mathematics Subject Classification.* Primary 68–01, 68Q25;
Secondary 05–01, 05C70, 05C80, 68R05, 90B80.

Library of Congress Cataloging-in-Publication Data
Knuth, Donald Ervin, 1938–
  [Mariages stables et leurs relations avec d'autres problèmes combinatoires. English]
  Stable marriage and its relation to other combinatorial problems : an introduction to the mathematical analysis of algorithms / Donald E. Knuth : [translated by Martin Goldstein].
    p. cm. — (CRM proceedings & lecture notes, ISSN 1065-8580 ; v. 10)
  Includes bibliographical references (p.   –   ) and index.
  ISBN 0-8218-0603-3 (alk. paper)
  1. Combinatorial analysis—Data processing. 2. Marriage theorem—Data processing.
I. Title. II. Series.
QA164.K5913   1996
511'.6—dc20   96-27510
                                                                                        CIP

English edition © 1997 by the American Mathematical Society. All rights reserved.
Reprinted with corrections 1997.
© 1981 Presses de l'Université de Montréal. Copyright for all languages.
This book was originally published in French under the title
"Mariages stables et leurs relations avec d'autres problèmes
combinatoires. Introduction à l'analyse mathématique des algorithmes"
Printed in the United States of America.

♾ The paper used in this book is acid-free and falls within the guidelines
established to ensure permanence and durability.
This volume was typeset using $\mathcal{AMS}$-LATEX,
the American Mathematical Society's TEX macro system,
and submitted to the American Mathematical Society in camera-ready
form by the Centre de Recherches Mathématiques.

10 9 8 7 6 5 4 3 2   02 01 00 99 98 97

... pour avoir appris,
la cuisine,
qui retient les petits maris,
qui s'débinent.

Contents

List of Figures

Foreword to the Original French Edition

These past ten years have seen the rapid development of an area that has been given the name Computer Science or Informatics. In my opinion, it would be more suitable to use the name "Algorithmics" for this discipline, whose principal object is not the study of computers themselves or information itself, but rather of the processes that treat information, or algorithms. Whatever it may be called, this area has turned out to be fascinating in more ways than one.

The goal of this work is to offer the reader an introduction to the analysis of algorithms based on examples, rather than going through the principal theoretical results. This type of presentation, I hope, will give the reader an idea of the methods used in this field and illustrate the close relation it has with a number of other branches of mathematics. The problem of stable marriage seems ideal in this context, since it does not require any prior knowledge of algorithmics and since it leads naturally to an illustration of the essential techniques of algorithmic analysis. This problem also allows me to show how interesting the analysis of algorithms can be in itself, in addition to its well-known practical importance.

The level of the discussion in this work is elementary and requires no prior experience either in analysis of algorithms or in marriage.

I must express my profound gratitude to Dr. André Aisenstadt whose generosity has permitted me to visit the Université de Montréal to give the series of lectures upon which this book is based. It was a pleasure to meet him on several occasions. I wish also to thank Professor Anatole Joffe for the honor bestowed on me by his invitation, and for his hospitality which made the visit as pleasant for my family as for me.

I regret not yet being able to master the French language. However, this deficiency rewarded me with the great pleasure of working closely with Pascale Rousseau and François Trochu, who have so carefully translated the text of these lectures given originally in English. I would like to thank, as well, Bernard Mont-Reynaud, one of my students, for his assistance in the preparation of the final text, and Micheline Marano and Johanne Marcoux for their excellent typing.

Donald E. Knuth
Stanford, California
May, 1976

N.B. On the occasion of the third printing of this book, I have corrected some typographic errors that slipped into the first and second printings. However, I am still waiting for solutions to the problems posed here, notably Problem 2!

D.E.K., February 1995

Translator's Note

There is a game played at parties where a phrase is quickly whispered from person to person and then announced to the group by the last recipient. The fun comes when the final phrase turns out to be nothing like the initial one. Translating Professor Knuth's lecture notes from French to English is like being the final player in the above game with some added complications. This time Donald Knuth "whispered" his lectures in English to the students named in the foreword who translated them and whispered them to me in French. I have translated these whisperings back to the original language and I am now announcing them to the world. I hope the outcome is less of a joke than the results of the party game.

To state this more formally, this book is a translation of the revised and corrected edition of *Mariages stables et leurs relations avec d'autres problèmes combinatoires* published in 1976 by *Les Presses de l'Université de Montréal*. It is based on the Aisenstadt Lectures given by Professor Donald E. Knuth at the Centre de recherches mathématiques. To keep the French flavor of the original version, we have not changed proper names and certain terms in the algorithms which are based on the French language.

I wish to express my gratitude to Louise Letendre for her excellent typing, to André Montpetit for his technical aid in producing the figures in the text and the format, and to the author for his careful editing of my translation. Finally my thanks go to nameless colleagues at the Université de Montréal who did not flee when I approached with my battered copy of the French version of this book to ask: "Now, how would you translate ... ?"

Martin Goldstein

LECTURE 1

Introduction, Definitions, and Examples

Let H and F be two finite sets of n elements. H is *the set of men* A_1, $A_2, \ldots, A_n$ and F is *the set of women* $a_1, a_2, \ldots, a_n$. A *matching* is a bijection of H onto F, i.e., a set of n monogamous marriages between the men and the women. In what follows, the two sets H and F play a symmetric role, and no special meaning should be attached to the fact that lower-case letters are used to denote the women.

Suppose that each man has an order of preference for the women and each woman an order of preference for the men. The list of preferences of the sexes is thus represented by two tables of n^2 elements.

A matching is *unstable* if a man A and a woman a, not married to each other, mutually prefer each other to their spouses. This "liaison dangereuse" occurs when:

- A is married to b;
- a is married to B;
- A prefers a to b;
- a prefers A to B.

(The opinions of b and B are irrelevant here.) A matching is *stable* if this situation does not occur.

EXAMPLE 1 (Marrying 4 men (A, B, C, D) to 4 women (a, b, c, d)).

Men	Order of preference				Women	Order of preference			
Anatole	c	b	d	a	antoinette	A	B	D	C
Barnabé	b	a	c	d	brigitte	C	A	D	B
Camille	b	d	a	c	cunégonde	C	B	D	A
Dominique	c	a	d	b	donatienne	B	A	C	D

Note that these preferences do not change with time.

The matching (Aa, Bb, Cc, Dd) is unstable since A and b prefer each other respectively (which is indicated by boldface A and b). Let us have Aa and Bb divorce, and A remarry b. Therefore, the only choice remaining for a and B is to marry each other. We obtain the matching (Ab, Ba, Cc, Dd), which is also unstable because of b and C.

Proceeding in this fashion until a stable matching is reached, we obtain the sequence

Aa	*Bb*	*Cc*	*Dd*	unstable
Ab	*Ba*	*Cc*	*Dd*	unstable
Ac	*Ba*	*Cb*	*Dd*	unstable
Ad	*Ba*	*Cb*	*Dc*	♡ stable ♡

To verify that we really have a stable matching, it suffices to produce a list of the wives that each man prefers to his own:

A prefers c then b
B prefers b
C has his first choice
D has his first choice

and the list of husbands that each woman prefers to her own:

a prefers A
b has her first choice
c prefers C, then B
d prefers B.

One cannot improve upon A's choice, since neither c nor b prefers him to her husband. Neither can we improve B's choice because b has already obtained her favorite partner. The matching considered is therefore stable.

We would arrive at the same result by taking the point of view of the women, i.e., by trying to improve the choices made by a, c, or d. We see that the men they prefer to their husbands do not share their feelings. ("C'est la vie.")

The procedure consisting of successive divorces, in the fashion indicated above, does not necessarily lead to a stable matching. To see this we look at the following example:

EXAMPLE 2.

Men's choice				**Women's choice**			
$A:$	b	a	c	$a:$	A	C	B
$B:$	arbitrary list			$b:$	C	A	B
$C:$	a	b	c	$c:$	arbitrary list		

The sequence of divorces and remarriages leads to a cycle, rather than terminating in a stable solution. Nevertheless, such solutions exist, for example (Ab, Bc, Ca) and (Aa, Bc, Cb).

$$
\begin{array}{ccc}
\boldsymbol{A}a & \boldsymbol{B}\boldsymbol{b} & Cc \\
A\boldsymbol{b} & Ba & \boldsymbol{C}c \\
Ac & B\boldsymbol{a} & \boldsymbol{C}\boldsymbol{b} \\
\boldsymbol{A}c & \boldsymbol{B}b & Ca \\
\boldsymbol{A}a & \boldsymbol{B}\boldsymbol{b} & Cc
\end{array}
$$

EXAMPLE 3 (Choices in a circular permutation).

Men's choice						Women's choice				
A :	a	b	c	d	e	a :	B	C	D	E	A
B :	b	c	d	e	a	b :	C	D	E	A	B
C :	c	d	e	a	b	c :	D	E	A	B	C
D :	d	e	a	b	c	d :	E	A	B	C	D
E :	e	a	b	c	d	e :	A	B	C	D	E

The five stable matchings are

$$
\begin{array}{ccccc}
Aa & Bb & Cc & Dd & Ee \\
Ab & Bc & Cd & De & Ea \\
Ac & Bd & Ce & Da & Eb \\
Ad & Be & Ca & Db & Ec \\
Ae & Ba & Cb & Dc & Ed
\end{array}
$$

In the first stable solution, each man is married to the woman of his choice while each woman is attached to her least favored man. The second stable solution improves the position of each woman by one rank while each man obtains his second choice. We continue similarly until the last solution, which corresponds to the optimum choice for the women and the worst for the men. This conflict of interest raises the question of defining a criterion for choosing a single stable solution. We might take the point of view of the women or that of the men.

We shall show that there exist only five stable solutions. A must be married to a, b, c, d or e. Suppose that a stable solution contains Aa. Then it must contain Ee. Indeed, if E were married to b, c or d, we see from the table that E would prefer a to his spouse and a would prefer E to A. Therefore the matching would be unstable. Thus we write

$$
Aa \implies Ee.
$$

Proceeding in this fashion, we obtain:

$$
Aa \implies Ee \implies Dd \implies Cc \implies Bb \implies Aa.
$$

For a stable solution to contain Ab, we must have Ee or Ea, but we have just seen that $Ee \implies Aa$. We thus obtain the sequence of implications

$$Ab \implies Ea \implies De \implies Cd \implies Bc \implies Ab.$$

Similarly if Ac appears in a stable solution, we have

$$Ac \implies Eb \implies \text{etc.}$$

This is an example where there are exactly n solutions. We can ask if n different solutions always exist. The existence of at least one stable solution will be established later. There exists in the literature an example where for $n = 8$, we have 9 stable solutions. If you are tempted to believe that there exist at most $2n$ stable solutions, or even n^2, the following example shows that there may be many more.

EXAMPLE 4 (n even).

Men's choice				Women's choice		
$1:$	①	2	$\ldots$	$1:$	$\ldots$	1
$2:$	②	1	$\ldots$	$2:$	$\ldots$	2
$3:$	3	④	$\ldots$	$3:$	$\ldots$	3
$4:$	4	③	$\ldots$	$4:$	$\ldots$	4
$\vdots$	$\vdots$	$\vdots$		$\vdots$		$\vdots$
$n-1:$	$(n-1)$	n	$\ldots$	$n-1:$	$\ldots$	$n-1$
$n:$	(n)	$n-1$	$\ldots$	$n:$	$\ldots$	n

Suppose that each pair of men (1 and 2, 3 and 4, ...) marries their first or second choice. The circles above illustrate such a marriage scheme, the number enclosed being that of the spouse. One can construct $2^{n/2}$ different schemes.

Each matching so obtained is stable. Indeed when two men marry their second choice, they cannot obtain their first choice because they are least preferred by the women in question.

One sees here that the systematic search for stable solutions can lead us to consider an exponential number of possible cases which, for n sufficiently large, takes too long, even for a computer.

Generalization to the problem of admitting n students to m universities

The k-th university admits n_k students where, without loss of generality, we suppose that $n_1 + n_2 + \cdots + n_m = n$. (If the number of students exceeds the possible number of admissions by the universities, we create a fictitious university, with the required number of admissions, which is the last choice of all the students. If the number of students is less than the number of places in the universities, we create fictitious students who are the last choice of all the universities. In this fashion we are led to the case of the equality $n_1 + n_2 + \cdots + n_m = n$.)

Each student has an order of preference for the universities and each university an order of preference for the students. We speak of *stable admission* when there does not exist a student A, not admitted to university a, such that A prefers a to his or her university and a prefers A to at least one of its students.

This problem reduces to that of stable matching if we replace the k-th university by n_k "places", each place having the same order of preference for the students.

In practice (for example the distribution of students among Iranian universities, or interns in American hospitals), this problem is formulated and resolved with the aid of a computer, by using an algorithm to determine a stable matching.

Generalization: Incomplete lists

Let us consider the problem of stable matchings when the men have not necessarily rated all the women and the women have not rated all the men.

We seek stable solutions with the extra condition that each person be married to someone appearing on his or her list of preferences (one can imagine that a person prefers to die rather than marry someone not on his or her list).

Consider for example the incomplete lists

$$
\begin{array}{llll} & & & \\ A: & a & & \\ B: & c & a & b \\ C: & c & a & \end{array}
\qquad
\begin{array}{llll} a: & C & A & B \\ b: & B & A & C \\ c: & A & B & C \end{array}
$$

The only possible matching is $(Aa, \boldsymbol{B}b, C\boldsymbol{c})$, but it is unstable because of B and c.

The existence theorem for stable matchings for complete lists does not therefore generalize to the case of incomplete lists.

NOTATION.

$$a \in A \iff a \text{ is on } A\text{'s list of preferences}$$
$$A \in a \iff A \text{ is on } a\text{'s list of preferences}$$
$$aAb \iff A \text{ prefers } a \text{ to } b \text{ or } b \notin A$$
$$AaB \iff a \text{ prefers } A \text{ to } b \text{ or } B \notin a$$

Thus $(A_1a_1, A_2a_2, \ldots, A_na_n)$ is a stable matching if and only if:

(i) $a_k \in A_k$ and $A_k \in a_k$ for $1 \le k \le n$;

(ii) there do not exist j and k such that $A_ja_kA_k$ and $a_kA_ja_j$.

One approach to resolving a matching problem of order n consists of reducing it to a problem of order $n-1$. One can easily show that a stable matching containing A_na_n exists if and only if $a_n \in A_n$, $A_n \in a_n$, and there exists a stable matching for the system B of order $n-1$ defined by the following properties:

$$b_i \in B_j \iff a_i \in A_j \text{ and not } (A_ja_nA_n \text{ and } a_nA_ja_i)$$
$$B_i \in b_j \iff A_i \in a_j \text{ and not } (a_jA_na_n \text{ and } A_na_jA_i)$$
$$b_iB_jb_k \iff a_iA_ja_k \text{ or } b_k \notin B_j$$
$$B_ib_jB_k \iff A_ia_jA_k \text{ or } B_k \notin b_j$$

for $1 \le i, j, k \le n-1$.

Conversion of incomplete lists to complete lists

One can produce a complete list from an incomplete list by adding a new man V, the widower, and a new woman v, the widow. The widower V will be a widow v's last choice, and v will be the woman V prefers least. The woman a_k will rate V last on her list of preferences (which is perhaps incomplete); then she will classify below V, in an arbitrary order, all the men that were not on her list.

Man A_k will place v last on his list of preferences (which is perhaps incomplete); then he will classify below v, in an arbitrary order, all the women that were not on his list.

An application of the preceding remark yields the following result:

THEOREM 1. *There exists a stable matching for the complete system such that V is married to v if and only if there exists a stable matching for the incomplete system.*

From results to be given in the next lecture, we can even show:

THEOREM 2. *If there exists a stable matching with V married to v for the complete system, then, for all stable matchings of this system, V is married to v.*

Consequently, to decide on the existence of a stable matching within the incomplete system, it suffices to obtain one stable solution for the complete system. If V is not married to v, then there does not exist a stable solution for the incomplete system. If V is married to v, then the stable solution without Vv is a stable solution of the incomplete system.

Exercises

1. Given the set of men A_1, A_2, ... , A_n, let a_k be the woman who represents the best "realistic" choice for the man A_k, that is, the woman he prefers among those with whom he might possibly attain a stable matching. All lists of preferences are complete. (We use here the existence, not yet established, of at least one stable matching.)

   (a) Prove that $j \neq k$ implies $a_j \neq a_k$.

   (b) Prove that $(A_1 a_1, \ldots, A_n a_n)$ is a stable matching.

2. Find all stable matchings for the choices

$$
\begin{array}{llllll}
A: & e & d & c & b & a \\
B: & a & e & d & c & b \\
C: & b & a & e & d & c \\
D: & c & b & a & e & d \\
E: & d & c & b & a & e
\end{array}
\qquad
\begin{array}{llllll}
a: & A & B & C & D & E \\
b: & B & C & D & E & A \\
c: & C & D & E & A & B \\
d: & D & E & A & B & C \\
e: & E & A & B & C & D
\end{array}
$$

3. (J. S. Hwang) Suppose that the preference matrix of the men is a Latin square (each column contains each of the women). Show that the matchings defined by the columns of this Latin square are all stable if and only if the preference matrix of the women is the dual Latin square (defined by the condition: a is the j-th choice of A if and only if A is the $(n + 1 - j)$-th choice of a, for all a and A).

LECTURE 2

Existence of a Stable Matching:
The Fundamental Algorithm

A fundamental algorithm allowing us to construct a stable matching will now be developed. This algorithm, in itself, constitutes a proof by construction of the existence of at least one stable solution. We saw in the first lecture that a random sequence of divorces does not always terminate in a stable matching. In the fundamental algorithm, the divorces are replaced by a number of engagements. The men in turn, one by one, play the role of suitor, making advances to the women, who accept or refuse according to their preference. We will see that this always yields a stable solution.

The algorithm uses three variables k, X, and x, and two constants n and Ω.

  n: number of men = number of women;
  k: number of (trial) couples already formed;
  X: suitor;
  x: woman toward whom the suitor makes advances
  Ω: (very undesirable) imaginary man.

Description of the algorithm

$k \leftarrow 0$; all the women are (temporarily) engaged to Ω;
while $k < n$ **do**
  begin $X \leftarrow (k+1)$-st man;
  while $X \neq \Omega$ **do**
    begin $x \leftarrow$ best choice remaining on X's list;
    if x prefers X to her fiancé **then**
      begin engage X and x;
      $X \leftarrow$ preceding fiancé of x
      end;
    if $X \neq \Omega$ **then** withdraw x from X's list
    end;
  $k \leftarrow k + 1$
  end;
celebrate n weddings.

For those who are not familiar with a language like ALGOL, we remark that $k \leftarrow 0$ indicates that the variable k takes the value 0, and that a series of instructions between the word **begin** and the corresponding word **end** constitutes a single sequence. This sequence is repeated when it appears in the program after the instruction **while** $\langle \cdots \rangle$ **do**, as long as the condition $\langle \cdots \rangle$ is satisfied.

EXAMPLE 1. To understand this algorithm, let us apply it to Example 1.

$$
\begin{array}{llll}
A: & c & b & d & a \\
B: & b & a & c & d \\
C: & b & d & a & c \\
D: & c & a & d & b
\end{array}
\qquad
\begin{array}{lllll}
a: & A & B & D & C & \Omega \\
b: & C & A & D & B & \Omega \\
c: & C & B & D & A & \Omega \\
d: & B & A & C & D & \Omega
\end{array}
$$

Note that we have added Ω at the end of the list of preferences of each woman.

At the start of the execution of the algorithm, k takes the value zero and each woman is engaged to Ω.

FIRST CYCLE. After we have verified that $k < n$, the variable X takes the value A (the first man). Next, we verify that $X \neq \Omega$ and then execute the instruction:

$$x \leftarrow \text{best choice remaining for } X,$$

that is,

$$x \leftarrow c.$$

At this point, the values taken by the three variables k, X, and x are as follows:

$$k = 0, \quad X = A, \quad x = c.$$

Since c prefers A to Ω, c and A become engaged. Thus, the value of X becomes Ω, the preceding fiancé of c. Since $X = \Omega$, we do not execute the command

withdraw x from X's list.

Instead, we go back to the instruction:

while $X \neq \Omega$ do.

Since $X = \Omega$, we exit the sequence of "advances" which begins with this instruction. The index k is thus increased ($k \leftarrow k + 1$).

SECOND CYCLE. Since $k < n$, we execute the instruction

$$X \leftarrow (k+1)\text{-st man},$$

and therefore X takes the value B; B becomes the new suitor. We verify that $X \neq \Omega$ and we execute the instruction

$$x \leftarrow \text{ best remaining choice for } X,$$

that is,

$$x \leftarrow b.$$

We thus have the situation:

$$k = 1, \quad X = B, \quad x = b.$$

The proposal of B being the first real advance made toward b, the couple Bb is formed. X becomes Ω and we increase k again.

THIRD CYCLE. This cycle is different from the preceding two. We verify that $k = 2 < n$. Thus $X \leftarrow C$. The best choice remaining on C's list is b. Thus $x \leftarrow b$ and

$$k = 2, \quad X = C, \quad x = b.$$

But b prefers C to her fiancé, B. Therefore we form the couple Cb and X becomes B, the former fiancé of b. After having verified that $X \neq \Omega$, we remove x from X's list (that is, b from B's list).

Since X has the value B, the instruction

while $X \neq \Omega$ do

again involves the execution of the sequence of "advances" with B as new suitor. The best remaining choice for B is a. Therefore $x \leftarrow a$ and

$$k = 2, \quad X = B, \quad x = a.$$

Woman a prefers B to Ω; anybody is better than he. Thus we form the couple Ba and X becomes her former fiancé, Ω. Again we leave the sequence of "advances" which begins with the instruction **while** $X \neq \Omega$ **do**; k is increased again.

FOURTH CYCLE. We verify that $k = 3 < n$. Thus $X \leftarrow D$ and the new situation is

$$k = 3, \quad X = D, \quad x = c.$$

Woman c prefers D to her current partner A. She changes partners and A becomes the new suitor, X. Then c is removed from A's list. The best choice remaining on A's list is b. Therefore

$$k = 3, \quad X = A, \quad x = b.$$

But b is already with C whom she prefers to A, so there is no change of partners. Miss b is removed from the list of A, who must make a new proposal to the best choice remaining on his list. We have

$$k = 3, \quad X = A, \quad x = d.$$

Woman d accepts the advances of A, since she prefers A to the undesirable Ω. Thus $X \leftarrow \Omega$ and we leave the sequence of "advances". Variable k is increased.

TERMINATION. Having reached $k = n = 4$, we terminate the iteration

while $k < n$ do.

We can now celebrate four marriages since, as we will show, the matching obtained is stable.

REMARK. One might ask why the algorithm contains two tests of $X \neq \Omega$ within the instructions

while $X \neq \Omega$ do ... ,

and

if $X \neq \Omega$ then ...

Some computer scientists believe it is better to make redundant tests than to use the command **goto**.

Proof of the algorithm

It is only recently (in the last five to ten years) that people have begun to prove results rigorously concerning algorithms. Usually intuition produces an algorithm, whose functioning is verified experimentally by computer. To prove the validity of an algorithm and evaluate its performance in a rigorous way, it is necessary to specify the meaning of the variables of the algorithm and to make clear their relationship at each step of the algorithm. Finally, induction is used on the steps of the computation.

We first note the following results which are valid during the execution of the algorithm.

POINT 1. *If woman a is removed from A's list, no stable matching can contain Aa.*

PROOF. When the algorithm effects the operation "remove x from X's list" with $x = a$ and $X = A$, two cases can arise:

(i) Man A has made advances to a, but she prefers her current partner B.

(ii) Woman a was engaged to A, but has just left him (after a more attractive proposal from B).

In both cases, a prefers B to A and B prefers a to all the other women remaining on his list.

If a stable matching contained Aa, then B would have to be married to someone he prefers to a. However no prospective partner preferable to a appears on B's list. Thus we reach a contradiction by induction on the algorithm (assuming that Point 1 is true at each of the steps preceding the one considered). □

POINT 2. *If A prefers a to his fiancée, it means that a has rejected him for another.* □

POINT 3. *Two women cannot be the fiancées of the same man (except if the man is Ω).* □

POINT 4. *A woman's situation never worsens throughout the course of the algorithm.* □

POINT 5. *The list of preferences of each man never becomes empty.*

PROOF. Otherwise this man would have been rejected by all the women because of Point 2. Therefore each woman has a "boyfriend" who is not Ω, because of Points 4 and 3. Thus there would exist at least n other men, besides A. Contradiction. □

Because of Point 5, the algorithm is *well defined*. Indeed all the other operations are clearly well defined; only Point 5 requires a proof.

The algorithm ends in a finite number of steps since after each proposition, either one of the lists of preference becomes shorter or variable k increases by one unit.

POINT 6. *The matching obtained is stable*

PROOF. If A is not married to a and if A prefers a to his spouse, it means that a has rejected him (Point 2) and is married to someone she prefers to him (Point 4). □

In fact, the algorithm gives *the optimal result for each man*. Each man has the best possible marriage: There is no stable matching in which he would have a spouse he prefers to the one he already has. (This is a consequence of Point 1. Note that we haven't used Point 1 to prove the validity of the algorithm, but only for this stronger property.)

Since the solution is characterized as being optimal for the men, it is independent of their numbering, even though the algorithm makes them play the role of suitor, one by one, in a particular order.

Furthermore, the result also constitutes the *worst solution for the women*. (In any stable matching, each women obtains a partner equal or superior to the one assigned to her by the algorithm.)

PROOF. Suppose that Aa is a marriage in the algorithm, but Ba and Ab are marriages in another stable solution where a prefers A to B. Thus A must prefer b to a, which contradicts the fact that Aa is the best solution for A.   □

The same algorithm can be used to obtain the best solution for the women (and the worst for the men). However, the women must make the advances.

Conflict of interest

The fact that "the best for the men is the worst for the women" is a special case of a much more general result.

THEOREM 3. *If one stable matching contains the couple Aa, and another contains couples Ab and Ba, then either*

$$bAa \quad and \quad AaB,$$

or

$$aAb \quad and \quad BaA.$$

(*In other words, every other stable matching is better for one of the spouses and worse for the other.*)

PROOF. By the definition of stability, the situations of A and a cannot both worsen in the second solution. Therefore it remains to show that they cannot improve for the two at the same time.

Let $A = X_0$, $a = x_0$, $b = x_1$, and suppose that A prefers b to a, that is $x_1 X_0 x_0$. Since the first stable matching represents a worse choice for X_0, it must be that x_1 has obtained a better choice.

Let X_1 be the husband of x_1 in the first stable matching. Thus $X_1 x_1 X_0$. Since the second matching yields a worse partner for x_1, it must be that X_1 has obtained a better choice. Let x_2 be X_1's wife in the second stable matching. Thus $x_2 X_1 x_1$, etc.

We obtain the sequence

$$X_0 x_0, \quad X_1 x_1, \quad X_2 x_2, \quad \ldots \quad \text{in the first stable matching,}$$
$$X_0 x_1, \quad X_1 x_2, \quad X_2 x_3, \quad \ldots \quad \text{in the second,}$$

where

$$x_{k+1} X_k x_k \text{ and } X_{k+1} x_{k+1} X_k \text{ for all } k \geq 0.$$

Since the number of persons is finite, there exist integers j and k, $j < k$, such that $X_j = X_k$. Let j be the smallest integer having this property and, for this j, let k be the smallest integer such that $X_j = X_k$ and $k > j$. We have $x_j = x_k$. Furthermore, $j = 0$ since otherwise $X_{k-1} x_k = X_{k-1} x_j$ would appear in the second matching as well as $X_{j-1}x_j$, from which $X_{j-1} = X_{k-1}$, contradicting the fact that j is the smallest integer with $X_j = X_k$. Thus $X_{k-1}x_0$ appears in the second matching. But $x_0 = a$. Thus $X_{k-1} = B$. Given that $X_k x_k X_{k-1}$, we have proved AaB. □

COROLLARY 1. *If a stable matching is at least as good as another from the point of view of each of the men, the second is at least as good as the first from the point of view of the women.*

Proof of the theorem stated in the first lecture

If there exists a stable matching containing Aa in which a is the last choice of A and A is the last choice of a, then *all* stable matchings contain Aa.

PROOF. The worst stable matching from the point of view of the women contains Aa. It is also the best solution for the men. □

Note that this theorem applies to lists completed by adding the elements V and v.

Analysis of the algorithm

In how many steps will the algorithm end? Let us see what happens with the help of the flowchart of Figure 2.1.

The index along an arrow indicates the total number of operations performed by the algorithm. The number of men or of women is n, which is a known constant. The total number of proposals of the suitors is denoted by N, and this number depends on the structure of the preference lists.

It is of interest to know the mean value of N when the preference lists are constructed at random. This will be the object of the next lectures. Another quantity of interest is the maximum that N can attain in the worst situation. That question is studied in the following exercises.

Exercises

1. Prove that at most one man obtains his last choice with the fundamental algorithm. [Consequence: If there exists a stable matching with two men (or more) obtaining their worst choice, then there exist at least two stable matchings.]
2. The result of Exercise 1 is the best possible: In certain cases, one man obtains his last choice and the other $n - 1$ obtain the next-to-last choice on their preference list. [In this case the total number of proposals is $n + (n - 1)(n - 1) = n^2 - n + 1$.]

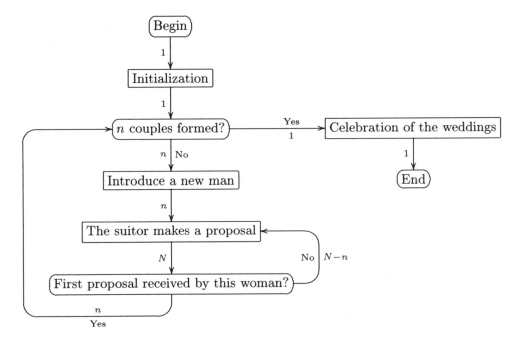

FIGURE 2.1. Flow of the fundamental algorithm.

Principle of Deferred Decisions: Coupon Collecting

In this lecture we estimate the mean number of proposals made in the course of the fundamental algorithm.

We first look at the following problem in order to introduce an important technique which is necessary for what follows.

The game of clock solitaire

We divide a deck of 52 cards into 13 stacks of 4 cards each, arranged face down as on the face of a clock (Figure 3.1).

Each hour indicates the number of a stack, except for:

- the stack at 1 o'clock, called "Ace",
- the stack at 11 o'clock, called "Jack",
- the stack at 12 o'clock, called "Queen",
- the stack at the center, called "King".

The game consists of turning cards over, one by one, according to the following rule:

- first we turn over the first card of the "King" stack;
- we place it at its appropriate spot on the clock (if it's a 7, for example, near the stack at 7 o'clock);

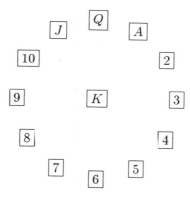

FIGURE 3.1. Tableau for clock solitaire.

- then we repeat this same procedure with the corresponding stack
  (we turn over the first card of the stack at 7 o'clock; if, for example, it's an ace, we place it near the "Ace" stack, etc.).

The game ends when we arrive at an empty stack. This occurs when
the fourth king is turned over. One wins if all cards are uncovered.

By a theorem in graph theory, there is a fast way to find out if
you win or lose. Everything depends on the bottom card of each stack
around the circle. Let us consider, for example, the distribution of
bottom cards shown in Figure 3.2.

The arrows indicate the positions to which each of the cards arranged around the circle must be moved. No arrow is drawn from the
"King" stack. We obtain a *directed graph* with 12 arcs and 13 vertices.

If this graph contains one or more cycles, then the player has lost;
he can not turn over all the cards. If there is no cycle, the player will
win, independent of the position of the other 40 cards.

The probability of winning is 1/13. There are two ways of showing
this:

(i) a difficult method consists of enumerating all the card distributions of the game such that no cycle appears among the last
cards in the stacks arranged on the circle.

(ii) a simpler method uses the *principle of deferred decisions*: "Don't
do today what you can put off till tomorrow". In particular, the
choice of the value of a turned-over card never has to be made
before the moment it is turned over.

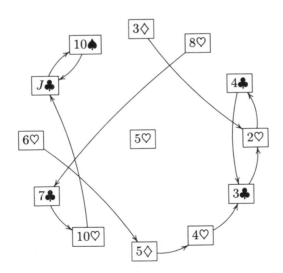

FIGURE 3.2. Linking the bottom cards.

We turn over the cards according to the rules of the game. Once the fourth king is turned over, we uncover the remaining cards, if any, by an arbitrary rule. (For example, turn over the remaining cards in the "Ace" stack, then the "Two" stack, etc.) We have now turned over all the cards in the game in a random order. When the initial distribution of the cards runs through the 52! permutations of the deck, the distribution obtained in turning over the cards in a particular order also runs through the 52! permutations of the deck. It is therefore easy to enumerate the winning distributions. The game is won if and only if the last card turned over is a king.

Study of the mean number of proposals

We seek the average number of proposals made by the successive suitors in the course of the algorithm to construct a stable matching. We suppose that the matrix of the women's preferences is given, but we assume that the matrix of the men is random. There are for each man $n!$ permutations of the n women, thus $(n!)^n$ equiprobable preference matrices. How do we determine the mean number of proposals in the algorithm?

We use the principle of deferred decisions. Instead of choosing in advance his list of preferences from among the $n!$ possibilities, each suitor makes up his mind at the last possible moment: For each new proposal, he chooses a woman at random to whom to address this proposal. This consists of choosing a card at random from the remaining cards.

However it remains difficult to calculate the mean number $E(N)$ of proposals in this way. It is necessary to simplify the problem.

Simplification of the problem

Let's suppose that the men are *amnesiacs*. When the moment arrives to make new advances toward a woman, they forget to which women they have already proposed.

This amounts to mixing the entire deck before the suitor picks a card to make a new proposal. The order in which these cards appear *for the first time* constitutes a random permutation.

The mean number $E(N)$ of proposals will be the mean number of proposals made by the amnesiac men, reduced by the mean number of redundant advances. In studying the simplified problem, we therefore obtain an upper bound for $E(N)$.

In the fundamental algorithm, the number of couples formed increases each time a woman receives her first proposal. The algorithm terminates as soon as each woman is proposed to at least once. With amnesiac men, the proposals form a random sequence that stops as soon as each woman is represented.

Consider the following example. Let A, B, C, D be four men and a, b, c, d four women whose lists of preferences are arbitrarily given as

$$
\begin{array}{cccc}
a: & C & A & D & B \\
b: & B & D & A & C \\
c: & C & A & B & D \\
d: & C & D & B & A
\end{array}
$$

Consider the sequence of proposals

$$d, b, d, b, c, c, b, c, d, a$$

which will serve for all successive suitors. We allow the fundamental algorithm to run, each suitor making advances to the available women in the order listed:

A proposes to d $\implies$ formation of the trial couple Ad,

B proposes to b $\implies$ formation of the trial couple Bb,

C proposes to d $\implies$ formation of the trial couple Cd

since d prefers C to A; A becomes the new suitor. A proposes to b; no new couple is formed since b prefers B to A.

A proposes to c $\implies$ formation of the trial couple Ac.

This amounts to constructing the following partial list of preferences:

$$
\begin{array}{l}
A: \quad d \quad b \quad c \\
B: \quad b \\
C: \quad d \\
D:
\end{array}
$$

We continue similarly until an advance is made to the last woman appearing on the list for the first time, namely a.

D proposes to c; c refuses.

D proposes to b; b refuses.

D proposes to c; c refuses again.

This approach is the first redundant proposal.

D proposes to d; d refuses.

D proposes to a $\implies$ formation of the trial couple Da.

The partial table of preference lists finally becomes

$$
\begin{aligned}
A &: \quad d \quad b \quad c \\
B &: \quad b \\
C &: \quad d \\
D &: \quad c \quad b \quad d \quad a
\end{aligned}
$$

Ten proposals have been made, one of which was redundant.

We note that it is useless to construct complete preference lists since, with the fundamental algorithm, one often obtains a stable matching before having referred to the entire preference matrix.

The length of the sequence necessary to obtain a stable matching in this way corresponds to the number of proposals made in the course of the algorithm. Therefore the mean length of such sequences corresponds to the *mean number* $E(N)$ *of proposals* (with amnesia). A sequence permits us to obtain a stable matching when all the women have appeared in it at least once. The calculation of the distribution of the lengths of such sequences constitutes the *coupon-collector's problem*.

Coupon collection

We suppose that there are n distinct coupons and that each time one buys a box of detergent one obtains a coupon at random. How many boxes must be bought, on average, to obtain all n coupons?

EXAMPLE 1. Here $n = 10$; we seek the number of steps to find all the digits 0, 1, 2, ... , 9 in the decimal expansions of π and e.

```
3  1  4  1  5  9  2  6  5  3  5  8  9  7  9  3
   2  3  8  4  6  2  6  4  3  3  8  3  2  7  9
   5  0
2  7  1  8  2  8  1  8  2  8  4  5  9  0  4  5
   2  3  5  3  6
```

We first specify the notation. Let p_k be the probability that exactly k boxes are necessary. We put

$$
\begin{aligned}
q_1 &= p_1 + p_2 + p_3 + p_4 + \cdots \\
q_2 &= \qquad\;\; p_2 + p_3 + p_4 + \cdots \\
q_3 &= \qquad\qquad\;\; p_3 + p_4 + \cdots \\
&\;\;\vdots \\
q_k &= \qquad\qquad\;\; p_k + p_{k+1} + \cdots
\end{aligned}
$$

Thus q_k is the probability that at least k boxes are necessary, and we have

$$q_1 = 1; \quad p_k = q_k - q_{k+1}.$$

The mean number of boxes used is, by definition,

$$p_1 + 2p_2 + 3p_3 + \cdots = q_1 + q_2 + q_3 + \cdots.$$

We can decompose the problem into n successive steps. At step m, $0 \leq m < n$, we're already in possession of m coupons and we are seeking the mean number of boxes that must be bought to obtain one more coupon. The probability of having to buy one or more boxes is evidently 1. (One has to buy at least one box to find a new coupon). Upon the purchase of a new box, the probability of finding one of the m coupons already in our possession is m/n. Thus at step m

$$q_1 = 1; \quad q_2 = \frac{m}{n}; \quad q_3 = \left(\frac{m}{n}\right)^2; \quad \cdots$$

$$q_1 + q_2 + q_3 + \cdots = 1 + \frac{m}{n} + \left(\frac{m}{n}\right)^2 + \cdots = \frac{n}{n-m}.$$

The mean number of boxes one must buy to obtain all n coupons is thus

$$\frac{n}{n-0} + \frac{n}{n-1} + \frac{n}{n-2} + \cdots + \frac{n}{n-(n-1)}$$

$$= n\left(1 + \frac{1}{2} + \frac{1}{3} + \cdots + \frac{1}{n}\right)$$

$$= nH_n$$

where H_n is the sum of the first n terms of the *harmonic series* $1 + 1/2 + 1/3 + \cdots$.

Conclusion

We have found an *upper bound for the mean number* $E(N)$ *of proposals* made in the course of the fundamental algorithm to obtain a stable matching. Indeed, the mean length of a random sequence of women such that the n women each appear at least once is equal to the mean number of boxes that must be bought to obtain the n coupons, namely nH_n. However

$$H_n = \ln n + \gamma + \frac{1}{2n} - \frac{1}{12n^2} + \varepsilon,$$

where $0 < \varepsilon < 1/120n^4$, γ is Euler's constant, and $\ln$ denotes the natural logarithm. Thus, for n big enough, the mean number $E(N)$ of proposals is asymptotic to $n \ln n$.

We can refine this result by considering a model with *partial amnesia*.

Partial amnesia

We now suppose that each man still has a very bad memory, but nevertheless he can remember the name of the woman who has just rejected him. This model gives a better upper bound for the mean number of proposals.

When proposals have been made to m women, $0 \le m < n$, the mean number of proposals that must be made before a man is accepted by an $(m+1)$-st woman can be calculated by slightly modifying the coupon-collection scheme:

$$q_1 = 1; \quad q_2 = \frac{m}{n}; \quad q_3 = \frac{m}{n}\frac{m-1}{n-1}; \quad q_4 = \frac{m}{n}\left(\frac{m-1}{n-1}\right)^2; \quad \cdots$$

$$q_1 + q_2 + q_3 + \cdots = 1 + \frac{m}{n}\left(1 + \frac{m-1}{n-1} + \left(\frac{m-1}{n-1}\right)^2 + \cdots\right)$$

$$= \frac{n}{n-m} - \frac{m}{n(n-m)}.$$

The probability of making one or more proposals is evidently $q_1 = 1$. The probability that the first advance be made to one of the m women is m/n. The probability that the other advances be made to one of the $m-1$ women among the $n-1$ remaining when the woman who has just rejected her suitor is removed from the list is $(m-1)/(n-1)$.

The mean number $\mathrm{E}(N)$ of proposals is easily calculated:

$$\mathrm{E}(N) = \sum_{m=0}^{n-1} \frac{n^2 - m}{n(n-m)} = \sum_{m=1}^{n} \frac{n^2 - (n-m)}{nm}$$

$$= (n-1)\left(\sum_{m=1}^{n} \frac{1}{m}\right) + \frac{1}{n}\left(\sum_{m=1}^{n} 1\right)$$

$$= (n-1)H_n + 1.$$

We thus obtain the following result:

THEOREM 4. *For every preference matrix of the women, the mean number of proposals made in the course of the algorithm ending in an optimal solution for the men is at most $(n-1)H_n + 1$.*

In this manner, each man makes at most about $\ln n$ proposals, on average. A woman receives approximately the same number of proposals. So she typically will have received one from a man who is $n/\ln n$ steps from the top of her list.

Exercises

1. Given the women's preference matrix

$$
\begin{array}{cccc}
a: & A & B & C \\
b: & B & C & A \\
c: & C & A & B
\end{array}
$$

consider the $216 = (3!)^3$ possible choices for the men. Show that there are exactly

> 48 matrices requiring 3 proposals,
>
> 72 matrices requiring 4 proposals,
>
> 60 matrices requiring 5 proposals,
>
> 30 matrices requiring 6 proposals,
>
> 6 matrices requiring 7 proposals.

[Thus the mean number of proposals is $4 + 5/12$; the upper bound is $2H_3 + 1 = 4 + 8/12$.]

2. Given the sequence of women *dbdbccbcda*, determine the probability that k redundant proposals are made, for each $k = 0, 1, 2, \ldots$, if the men are amnesiacs and the preference matrix of the women is random.

LECTURE 4

Theoretical Developments:
Application to the Shortest Path

Today we are going to examine the following points:

1. A quick review of the theory of discrete probability.
2. A study of the variance in the coupon-collector's problem.
3. The fundamental algorithm of stable matching: a study of the least favorable case.
4. Application to the calculation of the *shortest path on a graph*.

1. THEORY OF DISCRETE PROBABILITY

Generating functions

Let X be a *discrete random variable* taking values in the set $\mathbb{N}$ of nonnegative integers. Let p_k be the probability that $X = k$; thus $p_k \geq 0$ and $p_0 + p_1 + p_2 + \cdots = 1$.

The generating function for this probability mass function is the series

$$P(z) = p_0 + p_1 z + p_2 z^2 + \cdots = \sum_{k \geq 0} p_k z^k$$

We note that $P(1) = 1$.

The mean value $\mathrm{E}(X)$ and the variance $\mathrm{V}(X)$ of X can be calculated from the values of the derivative of the generating function at $z = 1$. We have

$$P(z) = p_0 + p_1 z + p_2 z^2 + \cdots = \sum_{k \geq 0} p_k z^k;$$

$$P'(z) = p_1 + 2p_2 z + 3p_3 z^2 + \cdots = \sum_{k \geq 0} k p_k z^{k-1};$$

$$P''(z) = 2p_2 + 6p_3 z + 12p_4 z^2 + \cdots = \sum_{k \geq 0} k(k-1) p_k z^{k-2}.$$

Let $E(X)$ be the mean value of the random variable X. We have

$$E(X) = \sum_{k \geq 0} k p_k = P'(1).$$

The variance $V(X)$ of the random variable X is defined as the mean of $\left(X - E(X)\right)^2$, that is

$$V(X) = \sum_{k \geq 0} \left(k - E(X)\right)^2 p_k$$

$$= \sum_{k \geq 0} k^2 p_k - 2 E(X) \sum_{k \geq 0} k p_k + E(X)^2 \sum_{k \geq 0} p_k$$

$$= \sum_{k \geq 0} k(k-1) p_k + \sum_{k \geq 0} k p_k - 2 E(X)^2 + E(X)^2$$

$$= P''(1) + P'(1) - P'(1)^2.$$

Significance of the variance

Let $\sigma = \sqrt{V(X)}$ be the standard deviation of the random variable X with the probability mass function p_k, $k \geq 0$.

If $m \geq 1$, the probability that $|X - E(X)| > m\sigma$ is less than $1/m^2$.

This result is independent of the probability mass function considered. It is called Chebyshev's inequality (a result due also to J. Bienaymé). Thus in at least 99% of the cases the following inequalities hold:

$$E(X) - 10\sigma \leq X \leq E(X) + 10\sigma.$$

PROOF. Let p be the probability that $|X - E(X)| > m\sigma$. Thus if $p > 0$,

$$V(X) = E\left(\left(X - E(X)\right)^2\right)$$

$$> p(m\sigma)^2 + (1-p) \cdot 0$$

$$= p m^2 V(X);$$

in other words, $p < 1/m^2$. $\square$

Independent random variables

Let Y be another random variable such that the probability that $Y = k$ is q_k. The corresponding generating function is

$$Q(z) = \sum_{k \geq 0} q_k z^k.$$

The random variables X and Y are independent if the probability that $X = j$ and $Y = k$ is $p_j q_k$, for all j and k.

If X and Y are independent, the probability that $X + Y = m$ is

$$r_m = p_0 q_m + p_1 q_{m-1} + \cdots + p_{m-1} q_1 + p_m q_0.$$

In this case, the generating function of the new random variable $X + Y$ is the product of the generating functions of X and Y; indeed

$$\sum_{m \geq 0} r_m z^m = \sum_{m \geq 0} \left(\sum_{\substack{j,k \geq 0 \\ j+k=m}} p_j q_k \right) z^m$$

$$= \left(\sum_{j \geq 0} p_j z^j \right) \left(\sum_{k \geq 0} q_k z^k \right)$$

$$= P(z) Q(z).$$

For each generating function $P(z)$, we write

$$e(P) = P'(1),$$
$$v(P) = P''(1) + P'(1) - P'(1)^2.$$

We can show that $P(1) = Q(1) = 1$ implies

$$e(PQ) = e(P) + e(Q),$$
$$v(PQ) = v(P) + v(Q).$$

PROOF. We have

$$(PQ)'(z) = P'(z)Q(z) + P(z)Q'(z),$$
$$(PQ)''(z) = P''(z)Q(z) + 2P'(z)Q'(z) + P(z)Q''(z).$$

Thus

$$(PQ)'(1) = P'(1) + Q'(1)$$
$$(PQ)''(1) + (PQ)'(1) - (PQ)'(1)^2 = P''(1) + 2P'(1)Q'(1) + Q''(1)$$
$$+ P'(1) + Q'(1) - P'(1)^2$$
$$- 2P'(1)Q'(1) - Q'(1)^2$$
$$= P''(1) + P'(1) - P'(1)^2$$
$$+ Q''(1) + Q'(1) - Q'(1)^2. \quad \square$$

Cumulative distribution

Let $q_k = p_k + p_{k+1} + \cdots$ be the probability that $X \geq k$. The generating function $Q(z) = q_0 + q_1 z + q_2 z^2 + \cdots$ does not correspond to any probability mass function, since $\sum q_k > 1$ except if $p_0 = 1$.

Since $p_k = q_k - q_{k+1}$ we have

$$P(z) = \sum_{k\geq 0}(q_k - q_{k+1})z^k$$

$$= \sum_{k\geq 0} q_k z^k - z^{-1}\sum_{k\geq 0} q_{k+1}z^{k+1}$$

$$= Q(z) - z^{-1}\big(Q(z) - Q(0)\big);$$

thus, since $Q(0) = 1$,

$$zP(z) = (z-1)Q(z) + 1.$$

By differentiating this relation, we obtain

$$P(z) + zP'(z) = Q(z) + (z-1)Q'(z),$$
$$2P'(z) + zP''(z) = 2Q'(z) + (z-1)Q''(z),$$

from which

(1) $$e(P) = Q(1) - 1,$$

(2) $$v(P) = 2Q'(1) + Q(1) - Q(1)^2.$$

2. VARIANCE IN THE COUPON-COLLECTOR'S PROBLEM

Let $P_m(z)$ be the generating function of the random variable: "the number of boxes one must buy to obtain an $(m+1)$-st coupon when one already has m of the n possibilities". And let $P(z)$ be the generating function for the number of boxes to collect all n coupons. The corresponding random variables being independent, we have

$$P(z) = P_0(z)P_1(z)\ldots P_{n-1}(z),$$

(3) $$e(P) = e(P_0) + e(P_1) + \cdots + e(P_{n-1}),$$

(4) $$v(P) = v(P_0) + v(P_1) + \cdots + v(P_{n-1}).$$

We find

$$Q_m(z) = 1 + z + \frac{m}{n}z^2 + \left(\frac{m}{n}\right)^2 z^3 + \cdots = 1 + \frac{nz}{n - mz},$$

$$P_m(z) = \frac{(n-m)z}{n - mz};$$

$$Q_m(1) = 1 + \frac{n}{n-m}, \quad Q'_m(1) = \frac{n^2}{(n-m)^2},$$

$$e(P_m) = \frac{n}{n-m}, \quad v(P_m) = \frac{mn}{(n-m)^2}.$$

Thus we obtain using (1), (2), (3), and (4),

$$e(P) = \sum_{m=1}^{n} \frac{n}{m} = nH_n = n\ln n + O(n);$$

$$v(P) = \sum_{m=1}^{n} \frac{n(n-m)}{m^2}$$

$$= n^2\left(1 + \frac{1}{2^2} + \cdots + \frac{1}{n^2}\right) - n\left(1 + \frac{1}{2} + \cdots + \frac{1}{n}\right)$$

$$= \frac{\pi^2}{6}n^2 - n\ln n + O(n).$$

The notation $O\big(f(n)\big)$ represents a function which when divided by $f(n)$ remains bounded for all sufficiently large n.

An improvement of Chebyshev's inequality

Since the standard deviation is $\sqrt{v(P)} = O(n)$, we deduce from Chebyshev's inequality that the number of proposals is almost always $nH_n + O(n)$ when the men make random advances with amnesia. However this inequality is applicable to arbitrary distributions, and in special cases we can often obtain stronger results.

Let us consider for example the distribution generated by $P_m(z)$. The mean is $n/(n-m)$ and the standard deviation $\sqrt{nm}/(n-m)$. Chebyshev's inequality thus tells us that the probability of buying more than $cn/(n-m)$ boxes before acquiring the $(m+1)$-st coupon is at most $m/n(c-1)^2$, for all $c > 1$. In fact we know that the probability of buying more than $cn/(n-m)$ boxes is

$$q_{1+cn/(n-m)} = \left(\frac{m}{n}\right)^{cn/(n-m)} \leq e^{-c},$$

since $e^x \geq 1 + x$ for all x and we may take $x = -(n-m)/n$. When c is large, this estimation is much better than that given by Chebyshev's inequality.

Let us now consider the coupon-collecting process in its entirety. The probability that more than cnH_n boxes must be purchased is at most equal to the probability that there is at least one step m in which more than $cn/(n-m)$ boxes are bought; this second probability is at most ne^{-c}. With the choice $c = C\ln n$, we obtain the following result:

THEOREM 5. *The probability of buying more than $CnH_n\ln n$ boxes, before completing the collection of n coupons, is at most $1/n^{C-1}$.*

Chebyshev's inequality, which only gives an upper bound of order $1/C^2(\log n)^4$, does not suffice to prove this theorem.

We have now shown that the number of proposals rarely exceeds nH_n, even with amnesiac men. But what happens in the least favorable case: How many proposals can we have? (Of course this question makes sense only in the absence of amnesia.)

3. FUNDAMENTAL ALGORITHM: STUDY OF THE LEAST FAVORABLE CASE

We now resolve the exercises of the second lecture.

One man at most reaches the end of his list, since at the moment he approaches the n-th woman the algorithm stops. (All the women have been proposed to.)

The $n-1$ other men each make at most $n-1$ proposals. The total number of proposals is thus at most $n + (n-1)^2 = n^2 - n + 1$.

Here is an example where this happens:

$$
\begin{array}{llllll}
A: & \textcircled{b} & \textcircled{c} & d & e & a \\
B: & \textcircled{c} & \textcircled{d} & e & b & a \\
C: & \textcircled{d} & e & b & c & a \\
D: & \textcircled{e} & b & c & d & a \\
E: & \textcircled{b} & c & d & e & a
\end{array}
\qquad
\begin{array}{llllll}
a: & \text{arbitrary list} \\
b: & B & C & D & E & A \\
c: & C & D & E & A & B \\
d: & D & E & A & B & C \\
e: & E & A & B & C & D
\end{array}
$$

- A proposes to b; formation of the trial couple Ab.
- B proposes to c; formation of the trial couple Bc.
- C proposes to d; formation of the trial couple Cd.
- D proposes to e; formation of the trial couple De.
- E proposes to b; formation of the trial couple Eb, since b prefers E to A;
- A becomes the new suitor.
- A proposes to c; formation of the trial couple Ac, since c prefers A to B;
- B becomes the new suitor.
- B proposes to d; formation of the trial couple Bd, since d prefers B to C;
- C becomes the new suitor,

and so forth ... until a receives her first proposal from A.

The algorithm stops when man A has reached the end of his list. Men B, C, D, and E have each made four proposals and are married to b, c, d, and e respectively. We see that these women have obtained their first choice. Whatever the preference list of a may be, the stable matching obtained is optimal for the women.

4. SHORTEST PATH IN A GRAPH

Consider n cities 1, 2, ..., n connected by roads of given length. Let L_{ij} be the length of the road leading from city i to city j. (If we cannot go directly from i to j, we denote $L_{ij} = \infty$.) It is not required that $L_{ij} = L_{ji}$, but we do insist that $L_{ij} \geq 0$.

An important algorithm, due to E. W. Dijkstra, gives the shortest distance from city 1 to each of the other cities. We note that this algorithm can be applied to other questions not related to road traffic: minimum time to perform a series of tasks, etc.

We first define the variables used in the course of the algorithm. (We will say that the distance between city 1 and another city j is known when we have determined the shortest path from 1 to j on the graph.)

A: set of cities for which the shortest distance to city 1 is known.
B: set of other cities.
d_i: shortest distance from 1 to i when one travels only through the cities of A.
k: city that moves from B to A.

Description of the algorithm

$d_1 \leftarrow 0$; $A \leftarrow \{1\}$; $B \leftarrow \{2,\ldots,n\}$;
for $i \in B$ **do** $d_i \leftarrow L_{1i}$;
while B is nonempty **do**
  begin
    $k \leftarrow$ city in B such that $d_k = \min\{d_i \mid i \in B\}$;
    $B \leftarrow B \setminus \{k\}$; $A \leftarrow A \cup \{k\}$;
    for $i \in B$ **do** $d_i \leftarrow \min(d_i, d_k + L_{ki})$
  end;

(The instruction $B \leftarrow B \setminus \{k\}$ removes the element k from the set B.)

EXAMPLE 1. Let's study the operation of the algorithm in the example of Figure 4.1.

The cities are numbered from 1 (Québec) to 10 (Montréal). The distances are indicated by a number along the path.

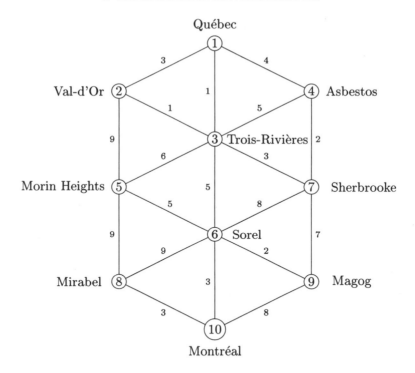

FIGURE 4.1. A shortest-path problem.

INITIALIZATION. No shortest path from Québec to another city is yet known:

$$d_1 \leftarrow 0; \quad A \leftarrow \{1\}; \quad B \leftarrow \{2, \ldots, 10\}.$$

The question is to determine the shortest path from Québec to the 9 cities of $B = \{2, 3, \ldots, 10\}$.

The only cities connected directly to Québec are Val-d'Or, Trois-Rivières, and Asbestos. Therefore $d_2 \leftarrow 3$, $d_3 \leftarrow 1$, $d_4 \leftarrow 4$, and $d_i \leftarrow \infty$ for the other cities.

FIRST STEP. Since B is not empty, we execute the sequence of instructions from **begin** to **end** (inclusive).

The closest city to Québec is Trois-Rivières. Thus

$$k \leftarrow 3, \quad B \leftarrow B \setminus \{3\}, \text{ and } A \leftarrow A \cup \{3\}.$$

(We remove 3 from the set B and add it to A because we know the shortest distance from Québec to city 3, as will be shown.)

For all the other cities of B (here for cities 2, 4, 5, ..., 10), we execute the instruction

$$\textbf{for } i \in B \textbf{ do } d_i \leftarrow \min(d_i, d_3 + L_{3i}),$$

that is we seek the shortest paths passing through a city in $A = \{1, 3\}$ which lead to the cities of B. We obtain the table

Cities of B
$d_2 = 2$
$d_4 = 4$
$d_5 = 7$
$d_6 = 6$
$d_7 = 4$

Cities of A
$d_1 = 0$
$d_3 = 1$

(For the other cities of B, d_i remains ∞.)

SECOND STEP. The set B is still nonempty: $B = \{2, 4, 5, \ldots, 10\}$. The city in B nearest to Québec, passing only through the cities of A, is the city ②; thus $k \leftarrow 2$, $B \leftarrow B \setminus \{2\}$, and $A \leftarrow A \cup \{2\}$. The new table obtained is

Cities of B
$d_4 = 4$
$d_5 = 7$
$d_6 = 6$
$d_7 = 4$

Cities of A
$d_1 = 0$
$d_2 = 2$
$d_3 = 1$

(The route to Morin Heights via Val d'Or is not shorter than that through Trois-Rivières, thus d_5 remains unchanged. For the cities of B not figuring in the table, $d_i = \infty$).

THIRD STEP. The set B is still nonempty: $B = \{4, 5, \ldots, 10\}$. The closest city in B to Québec (passing only through the cities of A) is now city ④ and $k \leftarrow 4$; $B \leftarrow B \setminus \{4\}$; $A \leftarrow A \cup \{4\}$.

We omit the table since there is no improvement in the distances d_i, $i \in B$. We have simply moved ④ into A.

FOURTH STEP. $k \leftarrow 7$; $B \leftarrow B \setminus \{7\}$; $A \leftarrow A \cup \{7\}$. We can now reach ⑨ ($d_9 = 11$) but no other distance d_i, $i \in B$, is improved.

FIFTH STEP. $k \leftarrow 6$; $B \leftarrow B \setminus \{6\}$; $A \leftarrow A \cup \{6\}$. We improve d_9 and ⑧ and ⑩ can be reached, as is indicated by the following table:

Cities of B
$d_5 = 7$
$d_8 = 15$
$d_9 = 8$
$d_{10} = 9$

Cities of A
$d_1 = 0$
$d_2 = 2$
$d_3 = 1$
$d_4 = 4$
$d_6 = 6$
$d_7 = 4$

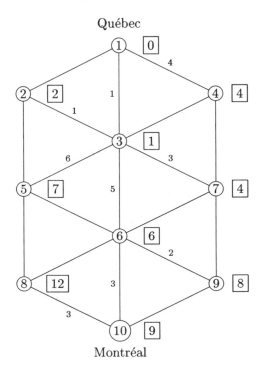

FIGURE 4.2. Distances from Québec.

We will now show that d_k is the shortest distance from Québec to the city k, at the moment that k moves from B to A. If there were a shorter path, it would necessarily go through a city in B other than k. But all the other cities of B are at least as far from Québec as k, and all the distances are positive or zero.

When the algorithm ends, the table of values d_i, $i = 1, 2, \dots, 10$, is shown in Figure 4.2.

The numbers in boxes are the shortest paths from Québec to the city in question. We have indicated the length of a path between two cities only if this route is traveled over in the optimal solution.

The mean execution time of the Dijkstra algorithm is difficult to determine exactly. But we can envision the following model:

> *For each city i, we randomly choose the cities j that will be connected to it on the graph, arranging these cities in the order of increasing L_{ij} (for each city i). We obtain a random permutation.*

For example, we choose the number of paths leading from each city according to a certain probability law. We then randomly determine the cities connected to each other by these paths. Let us suppose that these distances are independent identically distributed random variables.

Let us set up a preference list for each city i, by arranging the other cities in order of increasing L_{ij}. These lists are analogous to those of the algorithm for stable matching. One can thus study the mean execution time of the Dijkstra algorithm with the aid of results obtained in the study of the coupon-collector's problem.

Dijkstra's algorithm can be modified in such a way that the operation $d_i \leftarrow \min(d_i, d_k + L_{ki})$ is delayed until the moment it becomes necessary. We obtain an algorithm of the following form:

$A \leftarrow \{1\}$; $B \leftarrow \{2, \ldots, n\}$;
while B is nonempty **do**
    begin
    choose a city i in A whose list is nonempty;
    $k \leftarrow$ first city on i's list;
    erase k from i's list;
    if $k \in B$ **then** move k from B to A
    end;

Note the analogy with the stable-matching algorithm; the city k plays the role of the woman toward whom an approach is made. The criterion for refusing does not influence the analysis that we want to do. The important thing is that the variable k be randomly chosen. Thus the analysis of the coupon-collector's problem applies to the Dijkstra algorithm in the same way as to the problem of stable matching. The mean number of steps, in this second formulation of the Dijkstra algorithm, is at most nH_n. Therefore the algorithm, used n times to find the shortest distance from city i to city j, for all i and j, requires a mean number of steps of the order $O(n \log n)^2$.

Exercises

1. How many proposals are made in the stable-matching algorithm when all the men have exactly the same list of preferences for the women? (The choices of the women are arbitrary). Is the stable matching obtained unique?

2. How many proposals are made, on average, if the preference matrix of the men is

$$
\begin{array}{lcccc}
A: & a & b & c & d \\
B: & b & c & a & d \\
C: & c & a & b & d \\
D: & a & b & c & d
\end{array}
$$

and if that of the women is random?

3. Let X be a discrete random variable and p_k the probability that $X = k$. The Dirichlet generating function is

$$P(z) = \sum_{k \geq 1} \frac{p_k}{k^z}.$$

Note that $P(0) = 1$.

(a) Express $E(X)$, $V(X)$, and $E(\ln X)$ as functions of $P(z)$ and its derivatives.

(b) Let Y be a random variable independent of X. Its Dirichlet generating function is $Q(z)$. What is the Dirichlet generating function of the product $X \cdot Y$?

Express $E(XY)$ and $V(XY)$ as functions of $E(X)$, $E(Y)$, $V(X)$, and $V(Y)$.

LECTURE 5

Searching a Table by Hashing;
Mean Behavior of the Fundamental Algorithm

We have seen that the stable marriage problem has interesting relations with problems of coupon collecting and finding shortest paths. Now we will see that it also has a strong connection to a basic problem of information retrieval.

Hashing

Let x be a unit of information whose location in a table we seek. A standard method consists of arranging the information of the table in a certain order, then searching systematically for the unit of information x. (For example, looking up a name in the telephone directory.)

A more efficient technique, called hashing in English (and "hachage" or "adressage dispersé" in French), is useful inside a computer. If we have n available positions of memory, where n is larger than the number of data points, then certain sites are empty while others contain data. Suppose that the set of all possible data is extremely large, much bigger than the number n of positions of memory. (There are for example 26^6 6-letter names x, but only a small number of them will appear in any one run of the program.) We associate with each unit of information a number between 1 and n, which designates its position in the table.

Toward this end, we use a function $h(x)$ which associates with each name x a permutation of $\{1, 2, \ldots, n\}$, namely

$$x \to h(x) = h_1(x), h_2(x), \ldots, h_n(x).$$

In practice, it is possible to proceed in such a way that each permutation obtained is random (with the probability $1/n!$) and independent of the permutations associated with the other names in the table.

To find x in the table, we search the cells $h_1(x)$, $h_2(x)$, ... until we find x (successful search) or reach an empty cell (failed search). In the second case, we can place the name x in that empty cell. In this way we will be able to find it later.

EXAMPLE 1. Here $n = 9$ and the table contains six names whose permutations are

$$x_1 \to \mathbf{3}\ 2\ 7\ 8\ 6\ 9\ 1\ 5\ 4$$
$$x_2 \to \mathbf{1}\ 5\ 3\ 7\ 6\ 2\ 9\ 4\ 8$$
$$x_3 \to \mathbf{4}\ 8\ 1\ 2\ 5\ 9\ 6\ 3\ 7$$
$$x_4 \to \mathbf{1\ 5}\ 4\ 9\ 7\ 8\ 3\ 6\ 2$$
$$x_5 \to \mathbf{9}\ 8\ 6\ 4\ 1\ 3\ 5\ 7\ 2$$
$$x_6 \to \mathbf{4\ 1\ 7}\ 9\ 8\ 6\ 5\ 2\ 3$$

After having inserted the six names in the table, we have

1	2	3	4	5	6	7	8	9
x_2		x_1	x_3	x_4		x_6		x_5

If we are looking for x_4, for example, we will find it in two steps. We require only the number of steps in boldface in each permutation. If we seek a location for a name x_7 whose permutation is

$$x_7 \to 7\ 1\ 8 \dots,$$

we will find after three trials that x_7 is not recorded. After we place it in cell 8, we can find it in three steps.

Mean time to search for information

When we add an $(m + 1)$-st name to the table, what is the mean number of attempts before finding an empty cell in which to put it? (This gives the mean number of steps to find it later.)

We can formulate this question in an equivalent fashion: What is the mean number of steps in an unsuccessful search, when the table contains m names?

Let q_k be the probability that at least k steps are necessary. We have

$$q_1 = 1, \quad q_2 = \frac{m}{n}, \quad q_3 = \frac{m}{n}\frac{m-1}{n-1}, \quad q_4 = \frac{m}{n}\frac{m-1}{n-1}\frac{m-2}{n-2}, \quad \dots,$$

$$q_k = \frac{m!}{(m-k+1)!}\frac{(n-k+1)!}{n!} = \frac{\binom{n-k+1}{n-m}}{\binom{n}{m}}, \quad 1 \le k \le m+1,$$

where $\binom{n}{m} = n!/m!(n-m)!$ denotes a binomial coefficient. The mean number of steps necessary to place the $(m+1)$-st name is

$$q_1 + q_2 + q_3 + \cdots = \frac{\left(\binom{n}{n-m} + \binom{n-1}{n-m} + \cdots - \binom{n-m}{n-m}\right)}{\binom{n}{m}}$$

$$= \frac{\binom{n+1}{n-m+1}}{\binom{n}{m}} = \frac{n+1}{n+1-m}.$$

If the table is half full, then, on average, two steps are sufficient to establish that a name is not in the table.

The variance can be calculated in an analogous way. We obtain

$$\frac{(n+1)m(n-m)}{(n+1-m)^2(n+2-m)}.$$

When there are m names in the table, the mean number of steps to find a name (chosen at random) successfully is

$$\frac{1}{m}\left(\frac{n+1}{n+1} + \frac{n+1}{n} + \frac{n+1}{n-1} + \cdots + \frac{n+1}{n+2-m}\right) = \frac{n+1}{m}(H_{n+1} - H_{n+1-m}).$$

Let $\alpha = m/(n+1)$. Then the mean number of steps can be written as

$$\frac{1}{\alpha}\ln\left(\frac{1}{1-\alpha}\right) + O\left(\frac{1}{n}\right) \quad \text{when } n \to \infty,$$

since

$$H_{n+1} - H_{n+1-m} = \ln(n+1) - \ln(n+1-m) + O\left(\frac{1}{n}\right).$$

For example, when the memory is 90% full, the mean number of steps is only about $10\ln 10/9 \approx 2.56$, even if m and n are very large ($m = 900\,000$; $n = 1\,000\,000$).

Connection with the matching problem

We obtained an upper bound for the mean number of proposals in Lecture 3. In fact the true mean value can be somewhat smaller.

Let us calculate the mean number of proposals in the special case where the women all have the same preference list and the preferences of the men are random.

The total number of propositions does not change when one modifies the numbering of the men. Let suitor X be successively the first choice of the women, their second choice, etc. Each suitor thus proposes to women until a new woman is named. *Everything occurs exactly as in hashing.*

Let us return to the preceding example in supposing that there are 9 men and 9 women. Every woman prefers X_1 to X_2 to ... to X_9. The first 6 men $(X_1, X_2, \ldots, X_6)$ are already "encased": The cell that each occupies corresponds to the number of his girlfriend. The 7th man proposes to women 7 and 1, who refuse his advances (the cells 7 and 1 are occupied). Woman 8, named for the first time, will accept X_7.

The preceding results are therefore applicable. The mean number of proposals made by the $(m+1)$-st man is $(n+1)/(n+1-m)$. Thus the mean of the total number of propositions is:

$$(n+1)\left(\frac{1}{n+1} + \frac{1}{n} + \frac{1}{n-1} + \cdots + \frac{1}{2}\right) = 1 + (n+1)(H_n - 1)$$
$$= (n+1)H_n - n.$$

This number is less than the upper bound obtained for the model with partial amnesia by $n - 2H_n + 1$. We are led to make the following conjecture:

CONJECTURE. *The mean number of proposals, when the preferences of the women are fixed and those of the men are random, is always at least $(n+1)H_n - n$.*

In fact, given any preference matrix for the women, it seems reasonable to conjecture that, for all k, the probability that the number of proposals made by the men is $\geq k$ is minimal when the women have the same preference list. (This is evident for $k \leq n+2$; one can show it for $k = n+3$.)

Asymptotic value of the mean number of proposals in the fundamental algorithm

According to the above conjecture, we can expect the mean number of proposals to be close to the upper bound $(n-1)H_n + 1$, rather than our hypothetical lower bound $(n+1)H_n - n$. In fact, this can be shown and we arrive at the most important result in this series of lectures.

THEOREM 6. *Given random preference matrices for the men and the women, the mean number of proposals in the stable-matching algorithm is*

$$nH_n + O(\log n)^4.$$

Until now we have proceeded by enumeration (the combinatorial approach). In the proof of the fundamental theorem, a probabilistic approach (also called the Hungarian approach since it was developed by Erdös, Renyi, etc.) is used. One bounds the probability of the infrequent cases and in this way we need only consider the most frequent situations.

Let $M = 3n(\ln n)^2 - n$. We define the condition

CONDITION C. *There are at most M proposals.*

(Later we will see why this particular value of M was chosen.)

Let p be the probability that Condition C is satisfied. Suppose that the men are amnesiacs. When we run the fundamental algorithm, each woman classifies a man at the moment that he first proposes to her. At this moment she classifies the new suitor at random among the men who have already made advances to her. Let N be the number of true proposals and R be the number of redundant proposals. N and R are random variables but they are not independent; when N is large so is R. We know the mean and the variance of $N + R$ from the results of the coupon-collector's problem.

The probability, $1 - p$, that $N > M$ is less than the probability that $N + R > M$. From the theorem stated in the fourth lecture, we deduce that $1 - p = O(1/n^2)$. The mean number of true proposals $E(N)$ is thus

$$E(N) = p \cdot (E(N) \text{ when C is true}) + (1 - p) \cdot (E(N) \text{ when C is false})$$
$$\leq (E(N) \text{ when C is true}) + (1 - p)n^2$$
$$= (E(N) \text{ when C is true}) + O(1),$$

It remains to show that the mean number of proposals when C is true is $nH_n + O(\log n)^4$. This will follow from the fact that

$$(E(R) \text{ when C is true}) = O(\log n)^4,$$

since $E(N) + E(R) = nH_n$.

Let N_A and R_A be the number of proposals made by A. The distribution of these variables does not depend on A. Thus $E(N) = n E(N_A)$ and $E(R) = n E(R_A)$. We will show that the number of redundant proposals made by man A is $O\big((\log n)^4/n\big)$.

Let p_k be the probability that $N_A = k$ when Condition C is true. If $N_A = k$, the mean of R_A is

$$f(k) = \frac{0}{n} + \frac{1}{n-1} + \cdots + \frac{k-1}{n - (k-1)},$$

since the mean number of redundant proposals made between the m-th and $(m+1)$-st true proposal is

$$\frac{m}{n} + \left(\frac{m}{n}\right)^2 + \cdots = \frac{m}{n - m}.$$

The mean number $E(R_A)$ of redundant proposals made by A, when C is true, is thus:

$$E(R_A) = \sum_{k \geq 1} p_k f(k)$$
$$= (q_1 - q_2)f(1) + (q_2 - q_3)f(2) + (q_3 - q_4)f(3) + \cdots$$
$$= q_1 f(1) + q_2(f(2) - f(1)) + q_3(f(3) - f(2)) + \cdots$$
$$= q_1 \frac{0}{n} + q_2 \frac{1}{n-1} + q_3 \frac{2}{n-2} + \cdots + q_n \frac{n-1}{1},$$

where $q_k = p_k + p_{k+1} + \cdots$ and $q_{n+1} = 0$. In order to bound $E(R_A)$ from above, we will bound q_k, the probability that $N_A \geq k$ when C is true, by the probability that $N_A + R_A \geq k$ when C is true:

$$q_k \leq (1 - \varepsilon)^{k-1},$$

where $1 - \varepsilon$ is an upper bound for the probability that a proposal by A to a random woman a is rejected given all the proposals made to a by the other men. In other words, the probability that a accepts A is at least ε.

If the n women receive $m_1, m_2, \ldots, m_n$ proposals respectively from the other men during the running of the algorithm, the probability that A is accepted by a random woman is

$$\frac{1}{n}\left(\frac{1}{m_1 + 1} + \frac{1}{m_2 + 1} + \cdots + \frac{1}{m_n + 1}\right).$$

The minimum of the function $1/(m_1+1)+1/(m_2+1)+\cdots+1/(m_n+1)$, under the constraint $m_1 + m_2 + \cdots + m_n \leq M$, is attained when $m_i = M/n$ for $i = 1, 2, \ldots, n$. Thus the probability that an offer be accepted, given Condition C, is at least

$$\frac{1}{n}\left(\frac{n}{M+n} + \cdots + \frac{n}{M+n}\right) = \frac{n}{M+n} = \frac{1}{3(\ln n)^2},$$

which we take as the value of the parameter ε.

Recapitulation

Here is where things stand. We wish to prove the upper bound $E(R_A) = O((\log n)^4/n)$ and we know that

$$E(R_A) \leq \sum_{k=1}^{n} (1 - \varepsilon)^{k-1} \frac{k-1}{n+1-k}$$

where $\varepsilon = 1/\big(3(\ln n)^2\big)$. We finish the proof by using a classical technique:

$$
\begin{aligned}
\mathrm{E}(R_A) &\leq \sum_{1 \leq k \leq n/2} \frac{(1-\varepsilon)^{k-1}k}{n/2} + \sum_{n/2 < k \leq n} (1-\varepsilon)^{n/2}n \\
&\leq \frac{2}{n}\sum_{k \geq 1}(1-\varepsilon)^{k-1}k + \frac{n^2}{2}e^{-\varepsilon n/2} \\
&= \frac{2}{n\varepsilon^2} + \frac{n^2}{2}\exp\left(\frac{-n}{6(\ln n)^2}\right) = O\left(\frac{(\lg n)^4}{n}\right).
\end{aligned}
$$

The fundamental theorem is therefore proved. $\square$

Summary

Our analysis of the mean number of proposals in the algorithm for stable matching has permitted us to illustrate the fundamental techniques used in the analysis of algorithms:

1. Enumeration and finite summation.
2. Application of generating functions.
3. Combinatorial analysis of the least favorable case.
4. Probabilistic approach (also called Hungarian).

Final remark on the subject of hashing

In practice, it takes too long to calculate a truly random permutation $h_1(x)$, $h_2(x)$, ... , $h_n(x)$. We calculate $h(x)$ and $\delta(x)$, where $1 \leq h(x) \leq n$, $\delta(x)$ and n are relatively prime, and we use the permutation $h(x), h(x) + \delta(x), h(x) + 2\delta(x), \ldots$ (mod n). This technique is called "double hashing".

Even though double hashing uses only random arithmetic progressions, it has been shown recently by L. Guibas and E. Szemerédi that the asymptotic behavior is the same as if real random permutations were used. This difficult theorem was obtained by an elaborate application of the probabilistic approach. [A simpler proof was subsequently found by G. S. Lueker and M. Molodowitch, *More analysis of double hashing*, Combinatorica **13** (1993), 83–96.]

LECTURE 6

Implementing the Fundamental Algorithm

Now that we have analyzed the fundamental algorithm, we will write a *program* that can be given to a computer. To be efficient, this process of implementation requires the combining of diverse information structures to represent the data.

Inspection of the fundamental algorithm in the second lecture shows that the operations to translate are:

[1] $x \leftarrow$ best choice remaining on X's list.
[2] Does x prefer X to her fiancé?
[3] Engage X and x.
[4] Withdraw x from X's list.

In operation [1], we look for the first element x remaining on X's preference list. In operation [4], we remove an element from that list. Fortunately, this element is not chosen arbrarily; it is the first on the list. Let us choose here a structure such that these operations are easy to execute. We represent the preference matrix of the men by a two-dimensional table

$$\begin{pmatrix} HC[1,1] & HC[1,2] \ldots & HC[1,n] \\ \vdots & \vdots & \vdots \\ HC[n,1] & HC[n,2] \ldots & HC[n,n] \end{pmatrix}$$

where $HC[X, j]$ indicates the name of the j-th preference of X.

We also use a one-dimensional table $H[1], \ldots, H[n]$, where $H[X]$ is a pointer indicating the position of the first element remaining on X's list. When X is engaged, he is engaged to $HC[X, H[X]]$. The operations [1] and [4] are therefore written as follows:

[1] $x \leftarrow HC[X, H[X]]$;
[4] $H[X] \leftarrow H[X] + 1$.

Operation [2] consists of knowing if a new man X wouldn't be preferable to the current partner of woman x. We thus represent the women's lists differently from those of the men (otherwise the operation would require a search that is too long).

Let $P[x, X]$ be a number which serves to evaluate the preference of x for X; that is, $P[x, X] = n$ if X is her first choice, 1 if he is her last choice, and 0 if $X = \Omega$. Let $F[x]$ be the current partner of x. Then operation [2] is written:

[2] $P[x, X] > P\big[x, F[x]\big]$.

Finally operation [3] is easy to describe. We are therefore ready to give a concrete version of the algorithm.

```
        k, X, x, t: integer;
        HC: array[1:n, 1:n] of integer;
        P: array[1:n, 0:n] of integer;
        H, F: array[1:n] of integer;
 1.     assign values to HC and P;
 2.     for X ← 1 to n do H[X] ← 1;
 3.     k ← 0;
 4.     for x ← 1 to n do F[x] ← 0;
 5.     while k < n do
 6.         begin
 7.         X ← k + 1;
 8.         while X ≠ 0 do
 9.             begin
10.             x ← HC[X, H[X]];
11.             if P[x, X] > P[x, F[x]] then
12.                 begin
13.                 t ← F[x];
14.                 F[x] ← X;
15.                 X ← t
16.                 end;
17.             if X ≠ 0 then H[X] ← H[X] + 1
18.             end;
19.         k ← k + 1
20.         end;
21.     print the stable matching (HC[1, H[1]], ..., HC[n, H[n]]).
```

Improvements

Beginning with this first writing, we try to improve this new version of the algorithm by eliminating instructions that are not strictly necessary.

(a) Insert the operation $H[X] \leftarrow H[X] + 1$ before instruction 10. We thus avoid repeating the test: **if** $X \neq 0$. We must therefore change $H[X] \leftarrow 1$ on line 2 to $H[X] \leftarrow 0$.

(b) Introduce a new array $Q[x] \leftarrow P\big[x, F[x]\big]$, which avoids having to calculate $F[x]$ and $P\big[x, F[x]\big]$ for each new proposal made to woman x. Naturally $Q[x]$ must be modified each time that we change $F[x]$, but this does not happen often. (The operation $x \leftarrow HC[X, H[X]]$ could be modified in the same way, but this would not be an improvement.)

(c) If the compiler does not recognize the common expressions, we can change the sequence

$$H[X] \leftarrow H[X] + 1; \quad x \leftarrow HC[X, H[X]]$$

to

$$t \leftarrow H[X] + 1; \quad H[X] \leftarrow t; \quad x \leftarrow HC[X, t].$$

This improvement is not as important as the two preceding ones.

Initialization of the Table P

Let us suppose that the preference list of a woman is $(3, 2, 4, 1) = FC[x, \cdot]$. Then $P[x, \cdot]$ takes the values $(1, 3, 4, 2)$. For example, 3 is the first choice of x; thus $P[x, 3] = 4$. To do this assignment of values it suffices to use the following instructions:

for $x \leftarrow 1$ **to** n **do for** $k \leftarrow 1$ **to** n **do** $P[x, FC[x, k]] \leftarrow n + 1 - k$.

But to use memory more economically, we can let P share the same array as FC, by using the following algorithm (based on the cycle structure of each permutation):

```
for x ← 1 to n do
  begin
  for k ← 1 to n do FC[x, k] ← −FC[x, k];
  for k ← 1 to n do
    if FC[x, k] < 0 then
      begin
      j ← k;
      X ← −FC[x, k];
      while X ≠ k do
        begin
        t ← −FC[x, X];
        FC[x, X] ← n + 1 − j;
        j ← X;
        X ← t
        end;
      FC[x, k] ← n + 1 − j;
      end;
  end.
```

Now $P[x, X]$ appears in the FC array as $FC[x, X]$.

Here now is our program after the improvements (a) and (b) have been incorporated.

```
k, X, x, t: integer;
HC, P: array [1:n, 1:n] of integer;
H, F, Q: array[1:n, 1:n] of integer;
assign values to HC and P;
for X ← 1 to n do H[x] ← 0;
k ← 0;
for x ← 1 to n do F[x] ← Q[x] ← 0;
while k < n do
   begin
   X ← k + 1;
   while X ≠ 0 do
      begin
      H[X] ← H[X] + 1;
      x ← HC[X, H[X]];
      if P[x, X] > Q[x] then
         begin
         t ← F[x];
         F[x] ← X;
         Q[x] ← P[x, X];
         X ← t
         end
      end;
   k ← k + 1
   end;
print the stable matching (HC[1, H[1]], ..., HC[n, H[n]]).
```

Arranged marriage between A and a

Let us modify the program in order to find a stable solution in which A and a are married, and which is optimal for the men among these solutions (if they exist).

From the theory presented in the first two lectures, we use the fundamental algorithm with the $n-1$ men remaining when we exclude A, except for the following two situations:

[1] For each woman b whom A prefers to a, we forbid b to marry a man that she does not prefer to A.

[2] For each man B whom a prefers to A, we forbid B to marry a woman he does not prefer to a.

To eliminate situation [1], we start the algorithm by having A make an (insincere) proposal to the women b in situation [1]:

while $HC[A, H[A]] \neq a$ **do**
  begin $t \leftarrow HC[A, H[A]]$;
  $Q[t] \leftarrow P[t, A]$;
  $H[A] \leftarrow H[A] + 1$
  end

To eliminate situation [2], we interrupt the algorithm if a proposal from a man such as B is accepted by a. We are led to the following modified program:

k, X, x, t: integer;
HC, P: **array** [1:n, 1:n] **of** integer;
H, F, Q: **array** [1:n] **of** integer;
assign values to HC and P;
for $X \leftarrow 1$ **to** n **do** $H[X] \leftarrow 0$;
$k \leftarrow 0$;
for $x \leftarrow 1$ **to** n **do** $F[x] \leftarrow Q[x] \leftarrow 0$;
$H[A] \leftarrow 1$;
while $HC[A, H[A]] \neq a$ **do**
  begin $t \leftarrow HC[A, H[A]]$;
  $Q[t] \leftarrow P[t, A]$;
  $H[A] \leftarrow H[A] + 1$;
  end;
$F[a] \leftarrow A$;
$Q[a] \leftarrow P[a, A]$;
while $k < n$ **do**
  begin $X \leftarrow k + 1$;
  if $X \neq A$ **then while** $X \neq 0$ **do**
    begin $H[X] \leftarrow H[X] + 1$;
    if $H[X] > n$ **then goto** *done*;
    $x \leftarrow HC[X, H[x]]$;
    if $P[x, X] > Q[x]$ **then**
      begin $t \leftarrow F[x]$;
      $F[x] \leftarrow X$;
      $Q[x] \leftarrow P[x, X]$;
      $X \leftarrow t$;
      if $X = A$ **then goto** *done*
      end;
    end;
  $k \leftarrow k + 1$
  end;
print the stable matching $\big(HC[1, H[1]], \ldots, HC[n, H[n]]\big)$;
done:

Generalization to several arranged marriages

If we wish to find the optimal stable matching for the men such that the men in a subset S are married to women assigned to them, it suffices to slightly modify the order in which we introduce A to the woman a destined to be his wife.

for all $A \in S$ **do**
  begin
  $a \leftarrow$ spouse destined for A;
  $H[A] \leftarrow 1$;
  while $HC\big[A, H[A]\big] \neq a$ **do**
    begin
    $t \leftarrow HC\big[A, H[A]\big]$;
    if $Q[t] < P[t, A]$ **then**
      begin
      if $F[t] \neq 0$ **then goto** *done*;
      $Q[t] \leftarrow P[t, A]$;
      end;
    $H[A] \leftarrow H[A] + 1$;
    end;
  if $Q[a] > P[a, A]$ **then goto** *done*;
  $F[a] \leftarrow A$;
  $Q[a] \leftarrow P[a, A]$;
  end;

We also replace the instructions "**if** $X \neq A$" by "**if** $X \notin S$" and "**if** $X = A$" by "**if** $X \in S$".

Search for a fair stable matching

The different algorithms considered until now favor the men, and if we interchange the roles of men and women they would become favorable to the women. Such injustice is too shocking for the present day. Can we therefore find a solution that treats both sexes fairly?

We can enumerate all the stable solutions and choose the most satisfying matching according to certain criteria. This might take a long time if the number of stable matchings is large. We do not know in general if there exists a large or small number of solutions. It therefore seems preferable to use another method.

Let us choose for example the couple Aa at random. We run the modified algorithm which gives a stable matching containing Aa. If one exists, we repeat the same operation on a reduced problem of dimension $n - 1$, taking into account the fact that A is married to a. If there is no stable matching containing Aa, then we choose a couple other than Aa, and try again.

It is possible to develop this approach in such a way that the stable-matching algorithm is carried out at most n^2 times, hence with a mean number of steps of polynomial order (of the order $O(n^4)$), often $O(n^3 \log n)$.

Stan Selkow invented an algorithm less random than the preceding one, which leads to a *fair optimal solution* in the sense that it minimizes the regret of the most unhappy person. For a matching M, let us define the regret of a person by the distance of her (his) spouse from the top of her (his) preference list. Let $U(M)$ be the maximum regret of the $2n$ individuals. The idea is to construct a stable matching that minimizes $U(M)$. The algorithm is developed in the following way: The optimal solutions for the men and the women permit us to obtain upper and lower bounds for the regret of all the individuals in the stable matchings. We repeatedly form all marriages among individuals whose upper and lower bounds are equal. We thus obtain a reduction of the system. When we cannot further reduce the system, we choose an individual at random among those with the highest upper bounds. Without loss of generality, let us suppose that this individual is woman a; otherwise the roles of men and women are interchanged in the following description. Let A be a's worst choice from among the remaining men. Note that a is the best possible choice for A from the women who have not been eliminated. We remove a from A's list and we obtain a new optimal solution for the men. This raises the lower bound of A's regret and does the same for at least one other man. Simultaneously, the upper bound of a's regret, as well as that of at least one other woman, is diminished. It is clear that these iterative operations reduce the system to a stable matching.

Exhaustive search of stable matchings

We describe here a *recursive procedure "toutm"* which depends on a parameter, j, between 0 and n. At the beginning of the procedure, $(H[1], H[2], \ldots, H[n])$ denotes a stable matching. At the end, we return the variables of this array to their initial values. The algorithm prints all stable matchings $(H'[1], H'[2], \ldots, H'[n])$ such that

$$H'[k] \begin{cases} = H[k] & \text{for } 1 \leq k \leq j, \\ \geq H[k] & \text{for } j < k \leq n. \end{cases}$$

(No man makes a better choice, and the men from 1 to j do not change partners.)

To obtain all solutions to our problem, we first execute the fundamental algorithm. However when the time comes to print the optimal solution for the men, we use $toutm(0)$ in its place.

The program is written as follows:

```
        procedure toutm(j: integer):
        x, X, y, t: integer;
        SH, SF, SQ: array[1:n] of integer;
 1.     if j = n then print the stable matching (H[1], ..., H[n])
 2.     else begin for t ← 1 to n do
 3.             begin SH[t] ← H[t];
 4.             SF[t] ← F[t];
 5.             SQ[t] ← Q[t]
 6.             end
 7.     continue:   toutm(j + 1);
 8.     change:     X ← j + 1;
 9.                 y ← HC[X, H[X]]; (this partner will be forbidden to X)
10.     propose:    H[X] ← H[X] + 1;
11.                 if H[X] > n then goto finish;
12.                 x ← HC[X, H[X]];
13.                 if P[x, X] > Q[x] then
14.                     begin t ← F[x]; .
15.                     if t ≤ j then goto finish;
16.                     F[X] ← X;
17.                     Q[x] ← P[x, X];
18.                     if x = y then goto continue;
19.                     X ← t
20.                     end;
21.                 goto propose;
22.     finish:     for t ← 1 to n do
23.                     begin H[t] ← SH[t];
24.                     F[t] ← SF[t];
25.                     Q[t] ← SQ[t]
26.                     end
27.         end.
```

Here it is more difficult to prove the validity of the algorithm but the approach remains the same as before. By induction on the algorithm, we suppose that the executions of "$toutm(j + 1)$" are correct.

Then, we show that before performing the operation "$H[X] \leftarrow H[X] + 1$", all the desired stable matchings where X is paired with his choice $H[X]$ have already been printed. (Thus we find all stable matchings.)

Finally all the obtained matchings are stable. If man A prefers a to his partner, then a prefers her sweetheart to A since the quality of the women's companions improves in the course of executing the algorithm.

Passage to a non-recursive algorithm

This algorithm can be rendered non-recursive with the help of certain changes:

begin
x, X, y, t, j: integer;
SH, SF, SQ: **array**$[0{:}n,\ 1{:}n]$ **of** integer;
HC, P: **array**$[1{:}n,\ 1{:}n]$ **of** integer;
H, F, Q: **array**$[1{:}n]$ **of** integer;
$j \leftarrow 0$;
toutm: **if** $j = n$ **then** print the stable matching $\big(H[1], \ldots, H[n]\big)$;
else

    begin for $t \leftarrow 1$ **to** n **do**
      begin $SH[j,t] \leftarrow H[t]$;
      $SF[j,t] \leftarrow F[t]$;
      $SQ[j,t] \leftarrow Q[t]$;
      end;
    continue: $j \leftarrow j + 1$;
    goto *toutm*;
    change: (same sequence of instructions as before)
    propose: (same sequence of instructions as before)
    finish: **for** $t \leftarrow 1$ **to** n **do**
      begin $H[t] \leftarrow SH[j,t]$;
      $F[t] \leftarrow SF[j,t]$;
      $Q[t] \leftarrow SQ[j,t]$
      end;
    end;
$j \leftarrow j - 1$;
if $j \geq 0$ **then goto** *change*;
end.

One can show that the execution time of this program is $O(n^3 S)$ when there are S stable matchings. [An improved algorithm that requires only $O(n^2 + nS)$ steps was subsequently published by Dan Gusfield, *Three fast algorithms for four problems in stable marriage*, SIAM J. Comput. **16** (1987), 111–128.]

LECTURE 7

Research Problems

The following questions are far from the level of the famous problems presented by Hilbert in his celebrated lecture in Paris in 1900, but they seem to be worthy of interest and are likely to be solved in a finite time.

PROBLEM 1. Study the mean number of partner changes by women during the course of the fundamental algorithm. (A woman x changes her fiancé each time that we modify $F[x]$.)

Is this number much smaller than the mean number of proposals? It is rather natural to suppose that the number is of the order $n \log \log n$. Indeed a woman with random preferences always accepts the first offer, she accepts the second offer with probability $1/2$, ... , the k-th offer with probability $1/k$. If this woman receives k proposals, she changes fiancés, on average, $1 + 1/2 + 1/3 + \cdots + 1/k = H_k$ times. We know that a typical woman receives an average of H_n proposals; thus her mean number of changes of partner should be approximately $\log \log n$.

PROBLEM 2. In the course of the fifth lecture, we stated a conjectured lower bound on $\mathrm{E}(N)$, the mean number of proposals when the preference matrix of women is given. Is this conjecture true?

PROBLEM 3. If the preference matrix of the men is given and if that of the women is random, is the mean number of proposals maximum when all the men have the same order of preference for the women?

PROBLEM 4. Is there an efficient way to calculate exactly the mean number of proposals if the preference matrix of the women is given? In other words, can we find this number in an efficient way by studying the structure of the matrix of preferences of the women? Perhaps this number depends only on simple properties of the preference matrix, such as, for example, how many times two women rate two men differently. We would thus obtain an elegant solution of Exercise 1 following Lecture 3.

PROBLEM 5. Find preference matrices for n men and n women that maximize the number of stable matchings.

When $n = 4$, for example, the maximal number known is 10. This situation occurs for the following two preference matrices:

$$
\begin{array}{llllll}
A: & a & b & c & d \\
B: & b & a & d & c \\
C: & c & d & a & b \\
D: & d & c & b & a
\end{array}
\qquad
\begin{array}{llllll}
a: & D & C & B & A \\
b: & C & D & A & B \\
c: & B & A & D & C \\
d: & A & B & C & D
\end{array}
$$

where the stable matchings pair up A, B, C and D respectively with

$$
\begin{array}{llll}
a & b & c & d \\
b & a & c & d \\
a & b & d & c \\
b & a & d & c \\
b & d & a & c \\
c & a & d & b \\
c & d & a & b \\
c & d & b & a \\
d & c & a & b \\
d & c & b & a
\end{array}
$$

PROBLEM 6. Find a method to describe the structure of the set of stable matchings for given preference matrices, so that we can characterize the solutions without having to list them.

Remarks on this problem. The following theory, due to John Conway, shows the presence of such structures, but many questions remain unanswered.

THEOREM 7. *If $M = (Aa, Bb, \ldots, Zz)$ and $M' = (Aa', Bb', \ldots, Zz')$ are two stable matchings, then*

$$
M \vee M' = \left(A \max_A(a, a'), B \max_B(b, b'), \ldots, Z \max_Z(z, z') \right)
$$

is also a stable matching.
(Here $\max_A(a, a')$ represents A's preference from among $\{a, a'\}$.)

PROOF. We first show that $M \vee M'$ is a matching, i.e., the marriages are monogamous.

Suppose for example that

$$
a = \max_A(a, a') = \max_B(b, b') = b'.
$$

Then B is married to a in M'. Since M' is a stable matching and aAa', we have BaA. But M is stable. Therefore we also have bBa, which contradicts the fact that $\max_B(b, a) = a$.

Finally, the matching $M \vee M'$ is stable. Suppose, for example, that A and $\max_B(b, b') = b$ prefer to be together. Then A prefers b

to $\max_A(a, a')$, thus bAa. But AbB, contradicting the fact that M is stable. □

COROLLARY 1. *Under the hypotheses of the preceding theorem, the matching*

$$M \wedge M' = \left(A \min_A(a, a'), B \min_B(b, b'), \ldots, Z \min_Z(z, z')\right)$$

is also a stable matching.

PROOF. It suffices to interchange the men and the women and to apply the theorem of the second lecture. □

The operations $\vee$ ("max") and $\wedge$ ("min") are associative, commutative, idempotent, and distributive:

$$(M \wedge M') \wedge M'' = M \wedge (M' \wedge M''),$$
$$(M \wedge M') \vee M'' = (M \vee M'') \wedge (M' \vee M''),$$
$$M \wedge M = M,$$

etc.

There results:

COROLLARY 2. *Given any preference matrices for the men and the women, the set of stable matchings is a distributive lattice.*

The optimal solution for the men is the "max" of all the stable matchings, and the optimal solution for the women is the "min".

Let us study the lattice of stable matchings for some examples. We find in the literature the following example for 8 men and 8 women:

A:	e	g	a	b	f	h	d	c	a:	E	C	G	F	A	B	H	D
B:	b	c	g	e	d	a	h	f	b:	H	F	C	E	G	B	A	D
C:	h	e	a	d	f	b	c	g	c:	A	E	F	B	D	H	G	C
D:	c	b	g	d	a	f	h	e	d:	H	G	C	B	D	A	E	F
E:	g	b	e	a	c	f	h	d	e:	F	D	G	C	H	A	B	E
F:	a	f	g	e	h	d	b	c	f:	B	H	E	D	F	C	G	A
G:	b	e	g	f	c	d	h	a	g:	G	E	B	A	H	F	D	C
H:	c	h	d	e	g	b	f	a	h:	G	D	A	E	B	C	F	H

The nine stable matchings can be arranged according to the lattice below. (A matching is represented by the names of the partners of A, B, C, D, E, F, G, and H; the number above each name corresponds to the rank of the woman on her spouse's list.)

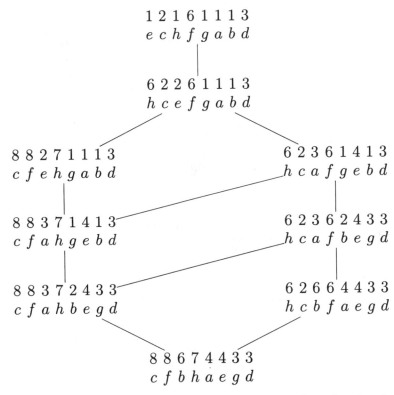

1 2 1 6 1 1 1 3
e c h f g a b d

6 2 2 6 1 1 1 3
h c e f g a b d

8 8 2 7 1 1 1 3
c f e h g a b d

6 2 3 6 1 4 1 3
h c a f g e b d

8 8 3 7 1 4 1 3
c f a h g e b d

6 2 3 6 2 4 3 3
h c a f b e g d

8 8 3 7 2 4 3 3
c f a h b e g d

6 2 6 6 4 4 3 3
h c b f a e g d

8 8 6 7 4 4 3 3
c f b h a e g d

We can represent this structure by the following distributive lattice:

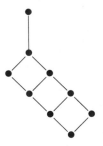

If we look over the lattice from top to bottom we notice that the preference rank that the men give to their wives is increasing.

The lattice representation of the ten stable matchings of problem 5 is

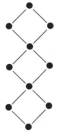

The lattice of the n stable matchings in example 3 of Lecture 1 is

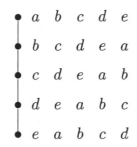

Here is an example where the lattice is a straight line, but the men do not all change partners from one matching to another:

$$
\begin{array}{ll}
A: & a \quad c \quad b \\
B: & b \quad c \quad a \\
C: & c \quad b \quad a
\end{array}
\qquad
\begin{array}{ll}
a: & B \quad A \quad C \\
b: & C \quad B \quad A \\
c: & A \quad B \quad C
\end{array}
$$

Here is an example where the lattice structure is "richer":

$$
\begin{array}{ll}
A: & a \quad b \quad c \quad d \quad e \quad f \\
B: & b \quad c \quad a \quad e \quad f \quad d \\
C: & c \quad a \quad b \quad f \quad d \quad e \\
D: & d \quad e \quad f \quad a \quad b \quad c \\
E: & e \quad f \quad d \quad b \quad c \quad a \\
F: & f \quad d \quad e \quad c \quad a \quad b
\end{array}
\qquad
\begin{array}{ll}
a: & E \quad F \quad D \quad B \quad C \quad A \\
b: & F \quad D \quad E \quad C \quad A \quad B \\
c: & D \quad E \quad F \quad A \quad B \quad C \\
d: & B \quad C \quad A \quad E \quad F \quad D \\
e: & C \quad A \quad B \quad F \quad D \quad E \\
f: & A \quad B \quad C \quad D \quad E \quad F
\end{array}
$$

We easily verify that

$$
\begin{array}{ll}
Aa \Longrightarrow Cc \Longrightarrow Bb \Longrightarrow Aa, & \qquad Dd \Longrightarrow Ff \Longrightarrow Ee \Longrightarrow Dd, \\
Ab \Longrightarrow Ca \Longrightarrow Bc \Longrightarrow Ab, & \qquad De \Longrightarrow Fd \Longrightarrow Ef \Longrightarrow De, \\
Af \Longrightarrow Ce \Longrightarrow Bd \Longrightarrow Af, & \qquad Dc \Longrightarrow Fb \Longrightarrow Ea \Longrightarrow Dc, \\
Ae \Longrightarrow Cd \Longrightarrow Bf \Longrightarrow Ae, & \qquad Db \Longrightarrow Fa \Longrightarrow Ec \Longrightarrow Db.
\end{array}
$$

Furthermore,

$$
\begin{array}{ll}
Ac \Longrightarrow Ff, Fd, \text{ or } Fe; & \qquad Ac, Ff \Longrightarrow Ee, Dd, Ba, Cb; \\
Ac, Fd \Longrightarrow Ef, De, Ba, Cb; & \qquad Ac, Fe \Longrightarrow Ba, Df, Cb, Ed.
\end{array}
$$

Thus

$$
Ac \Longrightarrow Ba \Longrightarrow Cb \Longrightarrow Ac,
$$

and the stable matchings have the following structure:

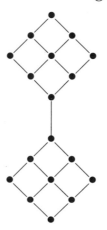

This can be interpreted as the *direct sum* of

and the two lattices

each of which is the *direct product* of the two lattices

Can one obtain all distributive lattices from suitable preference matrices? What special properties, if any, are shared by the lattices associated with dual Latin squares, as in Exercise 3 of Lecture 1? (The majority of examples above are of this type.)

One might try to generalize the problem in the following way: Instead of having totally ordered sets of men and women, each person could simply have a lattice order. But this does not work very well. Suppose that all the men rate the women according to the order given by the lattice and the women rate the men according to the same lattice:

There are two stable matchings (Aa, Bb, Cc, Dd) and (Aa, Bc, Cb, Dd). However $M \vee M' = (Aa, Ba, Ca, Dd)$ is no longer a matching and thus we do not have a lattice structure for the stable matchings.

Intersection of intervals

If we use the fundamental algorithm to find the optimal solution for the men and the optimal solution for the women, we obtain "intervals" of choice for each man and each woman. In all the stable matchings, each man is married to a woman from his interval. Similarly, each woman is married to a man from her interval.

Let us consider the example of 8 men and 8 women given before.

$$A : \quad e \quad g \quad a \quad b \quad f \quad h \quad d \quad c$$

$$B : \quad b \quad c \quad g \quad e \quad d \quad a \quad h \quad f$$

$$C : \quad h \quad e \quad a \quad d \quad f \quad b \quad c \quad g$$

$$D : \quad c \quad b \quad g \quad d \quad a \quad f \quad h \quad e$$

$$E : \quad g \quad b \quad e \quad a \quad c \quad f \quad h \quad d$$

$$F : \quad a \quad f \quad g \quad e \quad h \quad d \quad b \quad c$$

$$G : \quad b \quad e \quad g \quad f \quad c \quad d \quad h \quad a$$

$$H : \quad c \quad h \quad d \quad e \quad g \quad b \quad f \quad a$$

$$a: \quad \text{Ⓔ}_{\female} \quad C \quad G \quad \text{Ⓕ}^{\male} \quad A \quad B \quad H \quad D$$

$$b: \quad H \quad F \quad \text{Ⓒ}_{\female} \quad E \quad \text{Ⓖ}^{\male} \quad B \quad A \quad D$$

$$c: \quad \text{Ⓐ}_{\female} \quad E \quad F \quad \text{Ⓑ}^{\male} \quad D \quad H \quad G \quad C$$

$$d: \quad \text{Ⓗ}^{\male}_{\female} \quad G \quad C \quad B \quad D \quad A \quad E \quad F$$

$$e: \quad \text{Ⓕ}_{\female} \quad D \quad G \quad C \quad H \quad \text{Ⓐ}^{\male} \quad B \quad E$$

$$f: \quad \text{Ⓑ}_{\female} \quad H \quad E \quad \text{Ⓓ}^{\male} \quad F \quad C \quad G \quad A$$

$$g: \quad \text{Ⓖ}_{\female} \quad \text{Ⓔ}^{\male} \quad B \quad A \quad H \quad F \quad D \quad C$$

$$h: \quad G \quad \text{Ⓓ}_{\female} \quad A \quad E \quad B \quad \text{Ⓒ}^{\male} \quad F \quad H$$

Here $\male$ represents the optimal solution for the men, $\female$ the optimal solution for the women, "—" a marriage excluded because the woman is outside of the man's interval, and finally "/" a marriage excluded because the man is outside of the woman's interval.

For example, man A can not marry woman b because he is not in b's interval.

This observation considerably improves the speed of the exhaustive-search algorithm and the "fair" algorithms presented in the sixth lecture. However in certain exceptional cases such an observation has no effect (when the optimal solution for the men gives them their first choice while the women obtain their last choice, and when the optimal solution for the women gives them their first choice while the men obtain their last choice).

The intersection of intervals is a necessary condition for stability, but not a sufficient one. In the example above, there is no stable matching containing Bh, but the matching Ac, Bh, Ce, Df, Eb, Fa, Gg, Hd is in the intersection of the intervals.

PROBLEM 7. Given random preference matrices, find the asymptotic mean of the number of stable matchings. If $n = 2$, this number is $1 + 1/8$ and if $n = 3$ it is $1 + 1139/3888$. In the general case, this number is $n!\, p_n$ where p_n is the probability that $A_1 a_1$, $A_2 a_2$, $\ldots$, $A_n a_n$ is stable. We have

$$(1) \qquad p_n = \frac{1}{n!^n} \sum_{(x_{ij})} \prod_{1 \le j \le n} \frac{r_j!(n - 1 - r_j)!}{1 + c_j}$$

(2)
$$p_n = \sum_{(x_{ij})} \prod_{1 \le j \le n} \frac{(-1)^{r_j}}{(1+r_j)(1+c_j)}$$

where, in both cases, we sum over the $2^{n(n-1)}$ matrices (x_{ij}) where $x_{ij} = 0$ or 1 and $x_{ii} = 0$. Here $r_i = \sum_j x_{ij}$ is the sum of the i-th row and $c_j = \sum_i x_{ij}$ is the sum of the j-th column.

We obtain formula (1) by counting the preference matrices where $A_1 a_1, \ldots, A_n a_n$ is stable and $a_j A_i a_i \iff x_{ij} = 1$. To obtain formula (2), we put $x_{ij} = 1 \iff a_j A_i a_i$ and $A_i a_j A_j$ (unstable case). We add all the matrices, then subtract those with an unstable element, then add those with two unstable elements, etc. We thus obtain the probability that $A_1 a_1, \ldots, A_n a_n$ is stable. (It would be interesting to prove that (1) = (2) by algebraic manipulation.)

Here is another formula for p_n:

(3)
$$\int_0^1 \int_0^1 \cdots \int_0^1 dx_1 \, dx_2 \ldots dx_n \, dy_1 \, dy_2 \ldots dy_n \prod_{\substack{i,j=1 \\ i \ne j}}^n (1 - x_i y_j).$$

For example,

$$p_3 = \int_0^1 \int_0^1 \int_0^1 \int_0^1 \int_0^1 \int_0^1 da \, db \, dc \, dA \, dB \, dC \, (1 - At)$$
$$\times (1 - Ac)(1 - Ba)(1 - Bc)(1 - Ca)(1 - Cb).$$

PROOF. Notice that we can express the product

$$\prod_{\substack{i,j=1 \\ i \ne j}}^n (1 - x_i y_j) = \sum_{(x_{ij})} \prod_{j=1}^n (-x_j)^{r_j} y_j^{c_j}$$

and we obtain formula (2). □

The asymptotic behavior of these formulas seems rather delicate. We know that $n! \, p_n$ should be ≥ 1 since a stable matching always exists; but this does not follow in an obvious way from equations (1), (2), and (3).

PROBLEM 8. Given preference matrices for men and women, let us construct a "directed graph for the divorces". This graph will have $n!$ nodes, one for each permutation $p_1 \ldots p_n$ of $\{1, 2, \ldots, n\}$.

If $(A_1 a_{p_1}, \ldots, A_n a_{p_n})$ is an unstable matching in which A_i and a_{p_j} prefer each other to their respective spouses ($i \ne j$), then the graph of divorces contains an arc from node $(p_1, \ldots, p_n)$ to $(q_1, \ldots q_n)$, where $(q_1, \ldots, q_n)$ is the permutation $(p_1, \ldots, p_n)$ with p_i and p_j interchanged. The nodes from which no arc leaves represent the stable matchings.

Does there exist a path leaving from any point which leads to a stable matching? If so, is there one that is relatively small?

This directed graph might perhaps have some interesting properties related to the lattice structure of the stable matching.

PROBLEM 9. Does there exist an algorithm to find a stable matching where the number of operations increases less quickly than n^2 in the worst case? (We do not count the entry time of the preference matrices of the women and the men.)

PROBLEM 10. Does there exist an interesting connection between the problem of stable matching and the assignment problem? (Given a matrix a_{ij}, find a permutation maximizing $\sum_{j=1}^{n} a_{jp_j}$.)

PROBLEM 11. Can the stable-matching problem be generalized to three sets of objects (for example men, women, and dogs)?

PROBLEM 12 (A unisex problem: stability of roommates). Each individual from a set of $2n$ persons rates the other $2n - 1$ people according to a certain order of preference. Find an efficient algorithm (with execution time of polynomial order in the worst case) permitting us to obtain, if possible, a stable set of n couples. (Here stable means there are not two separated people who would each like to get back together.)

For example, we can verify that the system of preferences

$$
\begin{array}{llllllll}
A: & B & E & D & F & G & H & C \\
B: & C & F & A & G & H & E & D \\
C: & D & G & B & H & E & F & A \\
D: & A & H & C & E & F & G & B \\
E: & F & A & H & B & C & D & G \\
F: & G & B & E & C & D & A & H \\
G: & H & C & F & D & A & B & E \\
H: & E & D & G & A & B & C & F
\end{array}
$$

has only three stable solutions:

$$\{AB, CD, FG, HE\}, \{BC, DA, EF, GH\}, \text{ and } \{AE, BF, CG, DH\}.$$

Sometimes there is no stable solution:

$$
\begin{array}{lllll}
A: & B & C & D \\
B: & C & A & D \\
C: & A & B & D \\
D: & \text{arbitrary}
\end{array}
$$

(Each person associated with D would like to change.) Perhaps we can show that such a problem is NP-complete.

Résumé of the lectures

The problem of stable matchings has introduced us to a variety of methods in the analysis of algorithms. We have called upon notions from various areas of mathematics:

- Combinatorics,
- Probability Theory,
- Analysis,
- Algebra,

as well as subjects in computer science:

- Data Structures,
- Control Structures,
- Computational Complexity.

There remain many fascinating problems to explore.

Annotated Bibliography

Stable matchings

1. T. H. F. Brissenden, *Some derivations from the marriage bureau problem*, Math. Gaz. **58** (1974), 250–257. [Beware errors.]
2. Juan Bulnes and Jacobo Valdes, *Notes on the complexity of the stable marriage problem*, Stanford University, 1972, (unpublished). [The source of Exercises 1 and 2 of Lecture 2.]
3. D. Gale and L. S. Shapley, *College admissions and the stability of marriage*, Amer. Math. Monthly **69** (1962), 9–15. [The classic original paper.]
4. D. G. McVitie and L. B. Wilson, *Stable marriage assignment for unequal sets*, BIT **10** (1970), 295–309.
5. _____, *The application of the stable marriage assignment to university admissions*, Oper. Res. Quart. **21** (1970), 425–433.
6. _____, *The stable marriage problem*, Comm. ACM **14** (1971), 486–490. [Finding all stable marriages.]
7. L. B. Wilson, *An analysis of the stable marriage assignment algorithm*, BIT **12** (1972), 569–575. [The upper bound based on partial amnesia.]

Coupon-collector's problem

1. R. E. Greenwood, *Coupon collector's test for random digits*, Math. Tables Aids Comput. **9** (1955), 1–5.
2. D. E. Knuth, *Seminumerical Algorithms*: The Art of Computer Programming, vol. 2, Addison-Wesley, Reading, Mass., 1969, Section 3.3.2 B, Exercises 7–10.

Shortest-path problem

1. E. W. Dijkstra, *A note on two problems in connexion with graphs*, Numer. Math. **1** (1959), 269–271.
2. P. M. Spira, *A new algorithm for finding all shortest paths in a graph of positive arcs in average time* $O(n^2 \log^2 n)$, SIAM J. Comput. **2** (1973), 28–32.

Hashing

1. L. I. Guibas, *The analysis of hashing algorithms*, Ph.D. thesis, Stanford University, Computer Science Dept., 1976.
2. D. E. Knuth, *Sorting and Searching*: The Art of Computer Programming, vol. 3, Addison-Wesley, Reading, Mass., 1973, Section 6.4.
3. W. W. Peterson, *Addressing for random-access storage*, IBM J. Res. Develop. **1** (1957), 130–146.

4. J. D. Ullman, *A note on the efficiency of hashing functions*, J. Assoc. Comput. Mach. **19** (1972), 569–575.

Data structures and control structures

1. D. E. Knuth, *Fundamental Algorithms*: *The Art of Computer Programming*, vol. 1, Addison-Wesley, Reading, Mass., 1968.
2. _____, *Structured programming with go to statements*, Comput. Surveys **6** (1974), 261–301.

Algorithm analysis

1. D. E. Knuth, *The analysis of algorithms*, Actes du Congrès International des Mathématiciens (Nice, 1970), vol. 3, Gauthier-Villars, Paris, 1971, pp. 269–274.
2. _____, *Mathematical analysis of algorithms*, Information Processing 71 (Proc. IFIP Congress 1971, Ljubljana, 1971), vol. 1, North-Holland, Amsterdam, 1972, pp. 19–27.
3. R. Sedgewick, *Quicksort*, Ph.D. thesis, Stanford University, Computer Science Dept., 1975.

APPENDIX A

Later Developments

1. Dan Gusfield and Robert W. Irving, *The Stable Marriage Problem: Structure and Algorithms*, MIT Press Series in the Foundations of Computing, MIT Press, Cambridge, Mass., 1989, 240 pp.

[The theory of stable marriages has advanced greatly since these lectures were first presented, and significant new results continue to be discovered. This book presents an excellent summary of what was known in 1989, including the status of the 12 research problems of Lecture 7.]

2. Akihisa Tamura, *Transformation from arbitrary matchings to stable matchings*, J. Combin. Theory Ser. A **62** (1993), 310–323.

[Negative answer to Problem 8: Stability cannot always be achieved by a sequence of divorces.]

3. Fred Galvin, *The list chromatic index of a bipartite multigraph*, J. Combin. Theory Ser. B **63** (1995), 153–158.

[Uses the theory of stable marriages to resolve a famous conjecture of J. Dinitz.]

4. Donald E. Knuth, Rajeev Motwani, and Boris Pittel, *Stable husbands*, Random Structures Algorithms **1** (1990), 1–14.

[Shows that a woman has between $(\ln n)/2$ and $\ln n$ husbands, with probability 1, when preferences are random; simplifies the techniques of probabilistic analysis; uses the term "late binding" as a new name for the principle of deferred decisions.]

5. Donald E. Knuth, *An exact analysis of stable allocation*, J. Algorithms **20** (1996), 431–442.

[A related problem of matching traders to indivisible goods leads to interesting new relations between permutations. This paper includes further information about research problem 2, which remains the most fascinating unsolved problem about the fundamental algorithm for stable marriage.]

6.   Tomás Feder, *Stable networks and product graphs*, Mem. Amer. Math. Soc., vol. 555, 1995, 223 pp.

[This comprehensive monograph places the theory of stable marriage in the more general mathematical context of stable configurations in networks. Many elegant results are obtained, including new algorithms that yield solutions symmetrical with respect to men and women. For example, one can find a "minimum regret" stable marriage, in which the rank of the least favorable partner is as small as possible, in $O(n^2)$ steps; one can find an "egalitarian" stable marriage, in which the sum of ranks of all partners is minimized, in $O(n^3)$ steps.]

APPENDIX B

Solutions to Exercises

Exercises of Lecture 1

1. (a) If the stable matching $(A_j a_j, A_k b, \ldots)$ gives the best choice for A_j, and $(A_k a_j, A_j c, \ldots)$ is optimal for A_k, then $a_j A_k b$, from which we have $A_j a_j A_k$. On the other hand $a_j A_j c$, from which $A_k a_j A_j$. Contradiction.

   (b) Suppose that $a_k A_j a_j$ and $A_j a_k A_k$, an unstable situation. Let $(A_k a_k, A_j b, \ldots)$ be a stable matching that is optimal for A_k. Then $a_j A_j b$, by the definition of a_j (since $b \neq a_j$). By transitivity we have $a_k A_j b$, and thus $A_j a_k A_k$ contradicts the stability.

2. (Ae, Ba, Cb, Dc, Ed) and (Aa, Bb, Cc, Dd, Ee).

Exercises of Lecture 2

1. and 2. Cf. fourth lecture, third part.

Exercises of Lecture 3

2. Without loss of generality we can suppose that AdC. The probability of having 0, 1, 2, and 3 redundant proposals is respectively $1/9$, $7/18$, $1/4$, and $1/4$. For example, there is no redundant proposal when BbC, DbB, and CcD (probability: $1/12$) or when CbB, DbC, BcD, and CcB (probability: $1/36$).

Exercises of Lecture 4

1. Let $a\ b\ c\ \ldots$ be the common preference list. Woman a obtains her first choice, say A. The other men make their proposals to b who obtains her best choice other than A, say B. The remaining men make their proposals to c, who obtains her best choice other than A and B, and so on. The total number of proposals is $n + (n-1) + \cdots + 1 = n(n+1)/2$. The solution is unique since it is optimal for the men and the women. (The mean number of proposals would be less than nH_n if the women had made the advances.)

2. The order of proposals is always a, b, c, a, b, c, $\ldots$ until a man asks for d. The probability of having 7, 8, 9, 10, 11, 12, and 13 proposals

is respectively 1/4, 1/6, 2/9, 25/108, 1/12, 1/36, and 1/54. The mean is thus $8 + 8/9$. (See Problem 3 in Lecture 7.)

3. (a) $E(X) = P(-1)$, $V(X) = P(-2) - P(-1)^2$, $E(\ln X) = -P'(0)$.

(b) The generating function is $P(z)Q(z)$. $E(XY) = E(X)\,E(Y)$,
$V(XY) = \big(V(X) + E(X)^2\big)\big(V(Y) + E(Y)^2\big) - E(X)^2\,E(Y)^2$.

Index